The Rightful Place of Science:

Citizen Science

The Rightful Place of Science:
Citizen Science

Edited by
Darlene Cavalier &
Eric B. Kennedy

Contributors
Lily Bui
Darlene Cavalier
David Coil
Caren B. Cooper
Robert R. Dunn
Eric B. Kennedy
Bruce V. Lewenstein
Holly L. Menninger
Gwen Ottinger

Consortium for Science, Policy & Outcomes
Tempe, AZ and Washington, DC

The Rightful Place of Science series explores the complex inter-
actions among science, technology, politics, and the human
condition.

For information on The Rightful Place of Science series,
write to: Consortium for Science, Policy & Outcomes
PO Box 875603, Tempe, AZ 85287-5603
Or visit: http://www.cspo.org

Model citation for this volume:

Cavalier, D., and Kennedy, E. B., eds. 2016. *The Rightful Place of
Science: Citizen Science*. Tempe, AZ: Consortium for Science,
Policy & Outcomes.

ISBN: 0692694838

ISBN-13: 978-0692694831

LCCN: 2016908692

FIRST EDITION, JUNE 2016

CONTENTS

FOREWORD

Alex Soojung-Kim Pang

It's time to talk about citizen science. The movement to recharge the relationship between science and the public — to bring interested amateurs back into the world of scientific research, to collect and analyze data, to share computing resources, and to enable ordinary citizens to use the tools of science to shape policy, increase government accountability, and uncover corporate malfeasance — has been one of the most important and yet less noticed trends in the history of recent science. The publication of this volume is especially welcome as a guide to this movement.

For most of the history of science, the concept of "citizen science" would have been unnecessary. Almost all science was done by amateurs: people moved by passion, a love of learning, and the desire to improve the world. The distinction between amateurs and professionals evolved slowly, over the course of the nineteenth century, as graduate training programs turned out Ph.D.s in science and an expanding network of universities and government laboratories provided them with careers. Ironically, as science became more central to life in the modern world, it became more inaccessible: especially after World War II and the rise of Big Science, it became impossible for amateurs to make meaningful contributions to science, and harder to conduct well-informed public discussion of science and its social effects.

The citizen science movement doesn't seek to turn back the clock—no one who cares about science wants to undo centuries of progress!—but to use new technologies and media to give amateurs a place in the world of scientific research, and to bring scientific research more fully into the world. Just as blogs have enabled self-publishing, digital production tools support basement musicians, and computer-aided design and social media support the DIY and maker movements, ever-cheaper sensors, cloud computing, and a host of other tools make it possible for amateurs to collect high-quality data, to contribute to ongoing scientific projects, and to connect with professional scientists, politicians, and each other. In the process, they are expanding and redefining the place of scientific research in education and contemporary life.

In other words, if you were to imagine a movement that should attract the attention of scholars and researchers, that would showcase their tools, and give them a chance to observe all the issues that have concerned their fields for decades—about the credibility of evidence, scientific practice, the conduct of controversies, the role that gender and social forces play in shaping knowledge, and so on—you would create citizen science. If citizen science can advance the sciences, the study of citizen science will be a boon for scholars writ large.

The essays in this volume illustrate how scholars and practitioners alike can contribute to our understanding of citizen science, and offer some clues about how engagement with citizen science can improve scholarship as well. By way of conclusion, I'll point out just one. Writing about innovation sometimes requires being innovative about writing. The production of this volume escapes the problem that most academic work has when reporting on fast-moving subjects: by the time monographs about them finally appear, the subject has completely changed, and the opportunities for scholars to positively affect public policy and understanding have passed. Scholarly public engagement won't work if it takes place at the scholar's normal glacial pace, but reducing ourselves to talking heads with tweed coats isn't the solution either. Volumes like this one (and the other publications in *The Rightful Place of Science* series) represent an interesting experiment in making scholarly work accessible and timely.

Alex Pang, March 2016

PREFACE

Darlene Cavalier

Pietro Michelucci is a cognitive scientist at the Human Computation Institute who wants to find a cure for Alzheimer's disease. The most common form of dementia, Alzheimer's is marked by a number of physiological changes in the brain, including a reduction in the amount of blood flowing to the brain. Cornell University researchers believe that restoring that blood flow could slow or even cure Alzheimer's disease, but analyzing blood flow data is incredibly time-consuming and hard to automate.

Michelucci has a tool to solve the problem, but it's not a lab microscope, a brain scanner, or even a rack of computers. It's a crowd; actually two crowds, to be exact. The first is made up of 30,000 "dusters," people who participated in the stardust@home project in their spare time. Using an online virtual microscope, they sorted through a million images to find seven solitary particles of space dust captured by a satellite that was flown through the tail of a comet. The second crowd is the 200,000 people who played an online puzzle game called EyeWire to help trace the neural wiring of the human brain. Michelucci realized that the virtual microscope from stardust@home could be repurposed to help locate stalled blood vessels, and that the EyeWire puzzle game could be used to build a map of brain blood vessels. Combining the two would allow Alzheimer's researchers to see a 3D map of exactly where blood is and isn't flowing in the brain and speed up the research by a factor of 30. "By unleashing the power of the crowd, we can remove the analytic bottleneck and dramatically accelerate the Alzheimer's research," says Michelucci.

The people who join these crowds are called "citizen scientists." They're members of the general public who are participating in scientific research not as guinea pigs or funders, but by conducting experiments, making observations, collecting data, and engaging their minds to tasks beyond the reach of today's best computers. Not so long ago, "citizen scientist" sounded like a contradiction in terms. Science is remote, expensive, and requires a lab coat and a Ph.D.; amateurs might be able to appreciate science, just as they might appreciate opera or sports, but they couldn't contribute to it. Today, though, it's a short leap from supporting science to participating: enabled by technology and empowered by social change, curious laypeople are transforming the way science gets done. Some are students, others retired; some have ambitions to become scientists; others are driven by a love of nature or the challenge of an interesting problem; others want to use science to improve their neighborhoods or protect their environment. Citizen scientists work with professional scientists in academia or government, with grassroots organizations, or form their own social networks. They believe that research and discovery should be accessible and useful. (More than half of all basic research in the United States is federally funded, after all.) And it doesn't take a Ph.D. to grasp modern scientific problems like climate change, become involved in monitoring environmental conditions, or participate in policy discussions.

Satisfied Citizens

Necessity and innovation make it easier for people to get involved in serious science. The internet has dramatically reduced the cost and difficulty of sharing information and obtaining or using high-quality scientific instruments. Smartphones equipped with increasingly sensitive sensors, microphones, and cameras have become virtual laboratories, democratizing participation in science while extending the range and quality of data ordinary citizens can collect. Equipped with a smartphone and a few apps, citizen scientists can contribute to bird migration studies, track botanical phenophases, document glacier retreats and the impacts of fires on desert ecosystems, measure environmental noise and pollution, and even record earthquakes. (GPS and accurate clocks further assure that these observations have scientific value.) Social media can be used to coordinate group observations, publicize new findings, or share enthusiasm and excitement for science. Open access publications make it possible for citizen scientists to publish their findings and make their data discoverable to others who may want or need it.

Citizen scientists don't do research for a living; they practice science for personal satisfaction, to advance fields of research, or, sometimes, for self-preservation. Such was the case for the residents of Flint, Michigan, who knew something was wrong with their tap water. They shared their concerns with local officials, but little happened until they joined forces with Marc Edwards, a professor of civil engineering at Virginia Tech. Edwards led the study that proved there most certainly were high lead levels in their homes and, as his investigation would uncover, this was an issue the state scientists ignored. "They stepped forward as citizen scientists to explore what was happening to them and to their community, we provided some funding and the technical and analytical expertise, and they did all the work. I think that work speaks for itself," Edwards told the *Chronicle of Higher Education* in 2016.

Citizen scientists can extend the reach of existing instruments or observational networks. NASA's SMAP (Soil Moisture Active Passive) satellite orbits the globe every three days to measure soil moisture levels, but NASA also enlists citizen scientists to ground-truth satellite data by regularly reporting their local soil moisture levels. Together, the satellite- and citizen-generated data will be used to improve weather forecasts, detail water/energy/carbon cycles, monitor droughts, predict floods, and assist crop productivity. Similarly, thousands of people tweet their local snow depths and locations to #SnowTweets so that Richard Kelly, a cryosphere scientist from the University of Waterloo, can calibrate the accuracy of instruments on weather satellites. Why turn to humans? Clouds, for one thing, can block snow from being visible to satellites. "Citizen scientists' measurements, in many regions, match the snow cover model estimates," said Kelly. Across the country, more than a one and a half million amateur chemists and biologists monitor the quality of America's waterways.

Many of those water-monitoring citizen scientists organize into local chapters operating on $2,000 a year or less, and feed their findings into databases used by professional scientists and policymakers. They show that citizen scientists can make a material difference to research projects' bottom lines, and the speed at which science progresses. Presidential science advisor John Holdren estimated in September 2015 that time and labor donated by citizen scientists to biodiversity research had "an economic value of up to $2.5 billion per year." Amy Carton, former citizen science lead at Cancer Research UK, reported that her volunteers helped successful identify cancerous cells from drug trials by looking at slides in the online project Cell Slider. "It would have taken our researchers 18 months to do what citizen scientists did in just three months," she reported.

The Final Citizen Frontier

Thanks to more accessible technology, better data-gathering ability, easier coordination through the Internet and social media, and the growing track record of citizen science projects, opportunities to participate are becoming as diverse as science itself. Through these projects, citizen scientists are collaborating with professionals, starting projects on their own, conducting field studies, adding valuable local detail to research and even creating low-cost versions of expensive lab instruments to conduct their research. Their data are improving local decisions and policy-making. And their independence sometimes frees them to ask questions that lead science in new directions.

So what's next for citizen science? We may soon see the citizen science equivalent of Big Science or Revolutionary Science—discoveries and collaborations that bring together millions of people, and change the dynamics of innovation and research. Most citizen science projects today focus on data collection and analysis. This suits the needs of many researchers, and also builds on a growing interest among educators in teaching science in a more hands-on way. But Michelucci, the Alzheimer's researcher, envisions a day "when non-scientists contribute to all phases of the scientific process, including literature review, proposing new hypotheses, designing and running experiments, and data analysis and interpretation, and discovery." Of course, there are some fields that will always require years of study and incredible expertise, but "there is a growing body of evidence to support the validity of public participation in most facets of research," Michelucci says.

This field is already disrupting how science gets done. It's not difficult to imagine citizen science projects becoming part of high school and college curricula; local citizen science alliances influencing planning, zoning, and economic development plans; or citizens equipped with fitness monitors, smartphones, and drones contributing to global studies in biomedical science, sociology, ecology, and cosmology.

We have only begun to realize the vast potential of citizen science. This book aims to provide you with an introduction to citizen science, as represented by contributors from diverse fields. Whether you are a teacher, student, researcher, practitioner, policymaker, reporter, or a general-interest reader, we hope this book provides some insight, inspires you to read more about this topic, and perhaps even encourages you to get involved in citizen science projects that need your help.

INTRODUCTION

AN UNLIKELY JOURNEY INTO
CITIZEN SCIENCE

Darlene Cavalier

The American shad is Philadelphia's fish. Like the far more celebrated salmon, shad live their adult lives in cold, salty ocean waters and swim back to freshwater rivers and streams only to spawn. They're tasty like salmon, too, if bonier and less fleshy (the fish's Latin species name, *Alosa sapidissima,* means "most delicious fish"). Unlike salmon, though, shad can undertake their freshwater return migration several times in their lives — they are a most determined little fish. Shad were once so plentiful in the Philadelphia region that the Lenape Indians could hunt the fish in the Schuylkill and Delaware rivers with bows and arrows, and the shad industry provided the name for Fishtown, one of Philadelphia's archetypal neighborhoods. Philadelphians like me take pride in the shad's hardiness and history — they fed our country's Founding Fathers, after all, and were a dietary staple of city residents for generations.

By the mid-20th century, however, the people who lived along Philadelphia's rivers — many of whom depended on shad for their livelihoods — noticed that the shad were not migrating upriver as they had before.

They were being hampered by twin human-produced barriers, one chemical and the other physical. The industrialization that powered the city's prosperity had created a river system that was one of the most polluted in the country. Reportedly, the stink was so bad that military pilots were told to ignore the smell as they flew thousands of feet overhead. Meanwhile, as pollutants like phosphorous depleted oxygen levels in the rivers, a series of dams blocked migration routes; they established walls through which the shad couldn't pass and couldn't leap in their desperate attempts to reach their spawning grounds upstream. Fishermen and other locals did not know all the details at the time, but they observed declining fish numbers with great concern, knowing that the disappearance of the shad would affect their own economic and cultural survival.

Those citizens used what they did know about their environment, however, to guide their observations and inform their collection of data about local shad populations. With their findings, they were able to form hypotheses about the causes of the shad decline and communicate them to policymakers to encourage action in cleaning up the rivers. It was a process that sounds an awful lot like science and science-based policymaking.

I am inordinately fond of the shad, and perhaps I identify with the fish a little too closely. But how could I not? They are stubborn, persistent, maniacally focused creatures, and a legacy of a city I have called home for decades. It took a long, long time before the efforts of all those concerned citizens began to reverse the shad's fortunes—and only in very recent years has there been some real ground for optimism. Yet the shad's story provides a shining (albeit at times smelly) example of what can happen when non-professionals become involved in a scientific problem near and dear to their hearts. In some ways, their story mirrors that of my own journey and that of the field to which I have become dedicated: citizen science.

This book is intended to demonstrate the value and vitality of citizen science, and its terrific potential for involving many more everyday people in a dynamic and responsive scientific enterprise. This book is also addressed to people like me: those who, as young students, were not especially interested in dissecting frogs or working out physics problems, and had little desire to become professional researchers or engineers — but who, as adults, find themselves drawn to science, and more than a little curious as to how it shapes the world we live in. In some people, maybe, that interest shows itself as an itch to read about theories on the origins of the universe, or the search for unknown worlds or undiscovered species. Maybe it's a hunger to know more about what lies behind the ever-rising tide of technological wonders. Maybe the urge is for all things environmental: to know more about climate change or biodiversity or simply what kinds of birds are nesting in the backyard. Or perhaps it's a quest for greater clarity about the billions of federal tax dollars being spent on scientific research. There are a great number of us with such interests, and citizen science opens up a way for us all to become more involved in following our passions into the realms of research and policymaking.

In the diversity of projects described throughout this volume, the term "citizen science" encompasses a range of activities and involvement on the part of the public, a range large enough to include amateurs searching for hidden galaxies and middle school students documenting microbes culled from their belly buttons. Citizen scientists are often driven by an unending passion, whether to protect a species they care about, to speak up for people suffering from diseases or toxic exposures, or to watch over an ecosystem nearby. As Caren Cooper and

Bruce Lewenstein illustrate in Chapter 2, citizen science encompasses at least two main pursuits. One involves citizens voluntarily contributing observations and data to scientists, who then use this information in research. The other encompasses democratic participation in science and science policy, to ensure that it meets the needs and concerns of citizens. These are not mutually exclusive pursuits; indeed, one naturally engenders the other.

Because of this, citizen scientists can serve in a wide range of roles. Sometimes they are an educated volunteer researcher, collecting data, recording observations, and performing basic analyses. These roles can be especially useful on projects that are difficult to automate, where the human eye can make rapid work of complex problems. While these kinds of involvement have historically often been in one-time or context-specific roles, citizen scientists today can be involved in dozens of projects around the world. Sometimes, for instance, citizens are more active in designing and developing projects from the outset. For others, citizen science may mean a lifetime of government lobbying with science-based data. On other occasions, they're involved in research that would have been impossible a decade ago—like launching cube satellites into orbit.

All these components of citizen science increasingly overlap—that is, engaged citizens participating in scientific research desire a greater voice in how that research is conducted and what goals that research seeks to achieve. My own journey to citizen science certainly bears this out.

Swimming Upstream

I grew up in a blue-collar family, in a part of Pennsylvania where not many folks had the chance to go to college, or even the expectation that they should. I liked school well enough, and I got decent grades, but I was never particularly interested in my science classes. Our science teachers and the speakers they occasionally brought in—ostensibly to motivate us—seemed mostly to address only the handful of kids who were demonstrably smart and already science-oriented, leaving the rest of us to search for other passions to define us. In my case, those passions were the very non-scientifically taught disciplines of dancing and cheerleading, and I spent every waking classroom moment practicing routines under my desk.

All that practice paid off. After getting into Temple University's communications program, I made it onto the school's competitive cheerleading squad my freshman year. That provided me not only with excitement— I traveled the country and cheered at some thrilling games, including an NCAA playoff game at the University of Nevada Las Vegas—but also with an unlikely career start. In part because I needed to pay my way through college, in my senior year I landed a professional cheerleading gig with the Philadelphia 76ers, and for the next three years I got to share a court (or at least the sidelines) with Charles Barkley and Hersey Hawkins.

That was pretty much an evening job, though. During the day I worked for a company called Media Management, which performed administrative and marketing work for a variety of clients. One of those clients happened to be the popular science magazine *Discover*, and one of my tasks was to help with the newly inaugurated *Discover Magazine* Technology Awards program. The task at hand was chiefly organizational: I had to come up with suitable nominees for the various award categories, encourage them to apply, and then shepherd the submission of forms and supporting materials. But in the process I had to read through a mind-boggling variety of journals and magazines about everything from software design to medical research to environmentalism.

Not only did I learn a lot about recent scientific and technological developments, but I also interacted with the people at the heart of some truly amazing scientific research and cutting-edge technologies. Granted, my interaction with these titans was from a chair in the mailroom and often consisted simply of checking with them about missing information in their applications. But the innovators I spoke with — probably assuming I had some sort of influence on the $100,000 awards — were incredibly open and responsive to my requests for details about their work. All of which I found fascinating — as did my fellow cheerleaders, when I would talk to them in our dressing room about what I'd learned. That last fact may surprise most people, who do not readily associate cheerleaders with an interest in science. It did not in the least surprise me.

Before too long, my obvious interest helped me move out of the mailroom and into the classroom. I took over the Educator's Guide for *Discover*, reading the magazine cover to cover each month and translating the information into a form suitable for school use. I discovered (no pun intended) that the magazine, written for a general, nonprofessional audience and highlighting the most exciting developments in science and technology, was conceptually perfect for kids learning about science. It was perfect for my education, too: in learning more about how *Discover*'s writers and editors rearticulated complex material for broad understanding, and in how I could further explain it to teachers of young enthusiasts, I grew increasingly confident in navigating a once alien landscape.

Less than three years later, the magazine was bought by the Disney Company, and when I was hired by Disney and moved to their headquarters in New York City my responsibilities expanded considerably. They now included running the *Discover Magazine* Technology Awards, for which I'd been stuffing envelopes earlier. Eventually I became Senior Manager of Global Business Development for Walt Disney Publishing Worldwide, specializing in development and strategic marketing. This isn't meant to be a boast or even a recounting of my résumé. I mention it because my experiences at Disney opened my eyes to the fact that one of the important factors in the company's success was the partnerships and synergies they developed with others—a model that would eventually become enormously useful to me, and to the field of citizen science.

In my new role heading up the *Discover* Awards, I garnered a lot of corporate support for the program, and it became a significant annual event for both Disney and the scientific community. The Awards grew to become Disney's largest publishing event—the "Academy Awards of Science." There were thousands of applications and nominations, as well as annual two-week-long exhibitions and shows at Epcot Center. The role of celebrity judges grew impressively and included luminaries from *Apollo 11* astronaut Buzz Aldrin to magicians Penn & Teller, and from the famed physicist Freeman Dyson to the inimitable Ray Charles. Through the Awards and supporting science-themed roundtable discussions, I met F. Story Musgrave, the only astronaut to have flown missions on all five space shuttles and best known as the "fixer" of the Hubble Space Telescope; intriguingly, he was also a high school dropout who became a heart surgeon before becoming an astronaut. I met Marvin Minsky, co-founder of MIT's Media Lab and often referred to as the "father of artificial intelligence." I worked closely with astronaut Sally Ride, the first American woman to enter space, and Dean Kamen, the inventor of the Segway. Personally and professionally, it was a high point in my life, and interacting with some of the top scientific minds in America nurtured a deep love for science, a love that had been kindled just a few short years earlier.

It was inevitable, I suppose, that as that passion took greater hold of me, I began to wonder why it was so long in coming. What was it about my science classes in grade school that failed to inspire me in the way that conversations with professional scientists did? Perhaps it was the outdated "demonstration science" that passes for science education (which Robert Dunn and Holly Menninger eloquently critique in Chapter 3). Maybe, too, more insidious forces were at work: I had just assumed that science was intended for the geeky boys in my class—unaware of the subtle social pressures that girls receive, pushing us away from careers in science, technology, engineering, and math (STEM). Whatever the reason, I was grateful that I no longer saw science as something meant only for others. But I was the beneficiary of a series of truly fortuitous events. What about all the others like me who weren't so lucky?

It took a number of years with Disney before I had a conversation with an editor at *Discover* that changed my life. Over time my career had become very corporate—a daily march of business meetings and PowerPoint presentations—and I was telling the editor, Marc Zabludoff, how much I missed the work at *Discover*, educating the magazine's millions of readers about the ways science and technology impacted their lives and shaped the future. The editor interrupted my waxing nostalgic: "Do you think we really educate people? Or do we merely entertain them?"

He went on: "What do you think our readers can actually do with the information we give them? The opportunities for non-scientists to participate in anything having to do with science in a meaningful way are nil. People who aren't going to be scientists are excluded from the very start—after teaching the basics, science classes in schools are not geared toward kids who aren't planning to go into the sciences professionally. So we entertain people with the latest research and break-throughs, but there's not much the average person can do with that information, is there?"

"But isn't a scientifically literate population im-portant?" I objected. Don't we stress STEM education in school and fret that our students are falling behind other countries in science education? What is the point, if the scientifically literate can't engage with the research that impacts our future? How (as Lily Bui insightfully in-quires in Chapter 4) can media like *Discover* add value to the way citizens discover, assess, and even produce sci-entific information? Can't people like me, who aren't career scientists but are fascinated by science, participate meaningfully in the scientific enterprise—a huge and vital enterprise, I should emphasize, that's paid for in significant part by our tax dollars?

Marc was goading me, but he knew what I was really bemoaning—I had grown more comfortable with scien-tists, but I still felt I was little more than a tourist in the world of science. I wanted a place of my own. Claim one, he told me. "If you can figure out where you fit here, you'll figure this out for millions of people."

I took him up on his challenge. I applied to a gradu-ate program at the University of Pennsylvania and dove into science history and sociology. I was especially eager to learn how science policy worked, since policy is criti-cal for shaping what and how research is done in this country and because it seemed to offer an opening for non-scientists like me to get involved.

Through readings guided by Professor Susan Lindee, I started to understand how many lay people, like me, came to "find science." For many it was through the familiar path of activism—a response to a medical condition or disease outbreak or a local environmental concern. People who had a vested interest were quick to absorb technical information and take action. The environmentalists, notably, also organized communities to gather and share data and frequently called into question the ability of industry and government to place the interests of people first.

But at the time, 2004, the term "citizen science" (as coined by Cornell University's Rick Bonney) was still new. An internet search yielded very little of relevant interest. Apart from Cornell's Lab of Ornithology's small database of bird projects, there was no searchable listing of activities that allowed non-professionals the chance to be involved in scientific pursuits. That would soon change—new tools were being developed that would boost the citizen science movement enormously. Fuelled by the internet, data processing software, and the ubiquitous use of cell phones, it would become significantly easier to connect people to formal and informal research projects. Yet just a dozen years ago, it was still quite difficult to find these opportunities.

Among my more memorable projects in graduate school was a paper I wrote on the rise and fall of the U.S. Office of Technology Assessment (OTA), which provided Congress with objective analyses of important issues in science and technology from 1972 to 1995. Throughout six administrations, both Republican and Democratic, this small agency provided Congress with unbiased information about a host of critical scientific, technological, and environmental issues—from acid rain to radioactive-waste storage, from solar power to AIDS prevention—before being shut down during the days of Newt Gingrich's reign as Speaker of the House. I probably read every OTA report the office produced during its 23 years of existence, and many of the recommendations for reopening it after Congress shut it down. The OTA proved to be a very influential creation, and a number of other countries, especially in Europe, modeled their own technology assessment institutions on it. Yet it was defunded here, despite much critical acclaim for its work, and without any true input from the public on its worthiness.

For my master's thesis I expanded on the issues raised by the demise of the OTA, exploring how average citizens can engage with the complexities of national science policy, and how they can voice their knowledge and values on an equal footing with acknowledged experts. It was then that I first truly encountered that remarkable group of people known as citizen scientists and the barriers they were trying to tear down.

Through their grassroots, bottom-up efforts, they were aiding research by tagging butterflies, monitoring water health, keeping an eye on bird migratory patterns, and looking for new galaxies. But when it came to engaging in policymaking decisions, they were shut out. The forces against them were considerable, coming from politics and industry. But there was also strong resistance from the scientists themselves.

Scientists and other experts seemed to fear that the lay public, largely lacking formal science education, could not grasp technical concepts as they relate to policy. By and large, they concluded that unless people possessed credentialed scientific expertise, they should be excluded from any discussion of how research into such topics as, say, synthetic biology, biomedicine, alternative energy, or climate change should be funded or applied.

To my mind, this was wrongheaded, and not just because a democratic government is supposed to represent the will of its citizens. I thought it incredibly important for all interested people to be involved in such decision making because we live in a society in which science and technology are major drivers of social and economic change—that's why we invest huge sums of money in them. The changes brought about by science and technology can be responsive to society's needs and meet the enormous challenges confronting all of us. Opening up the process of how scientific resources are allocated and assessed, or at the very least making these processes more transparent, struck me as an obvious win-win: citizens would be more knowledgeable about the science being done in their name, scientists and policymakers would be able to better anticipate challenges and do some risk assessment before they rolled out new policies, and the societal benefits of our research and development investments would be vastly improved.

I had expected the resistance I experienced from politicians I met with in those days. Newt Gingrich, for example, offered the usual talking points for the demise of the OTA under his watch: that it merely represented bloated government, that it couldn't offer neutral assessments, and that if he needed a technological analysis he could call the appropriate people himself. He viewed the OTA as simply a unidirectional source of potentially biased analysis, rather than as a way of engaging constituents in science policy for the edification of both policymakers and the public.

But I was disheartened when in years to come I encountered a similar lack of understanding by scientists. Speaking with me at an event on citizen engagement, for example, was an official representative from a professional science association, who was ostensibly presenting in support of such lay participation. Before the event, though, she leaned over to me to say: "By the way, you're completely misguided." She elaborated, arguing that there was already a system in place for citizen input once a bill has been posted, called the "public consultation period," when people could provide comments to the bill. In her eyes, there was no need for upstream public engagement of the sort that I advocated, especially with a population that isn't particularly scientifically literate. Never mind the fact that research by the University of Michigan's Jon Miller found the scientific literacy of U.S. adults is relatively high compared to other developed nations!

Changing the Current

That moment was really the start of my citizen science advocacy, and it has shaped all my activities since. Ten years ago, I started pushing to reopen the OTA, which I thought had the most potential to bring together the public and scientists in shaping science policy. I supported efforts by professional science organizations such as the Union of Concerned Scientists and members of Congress, including Rush Holt, then a Representative from New Jersey (and current CEO of the American Association for the Advancement of Science). However, unlike the stated aims of such OTA supporters, my goal was to embed mechanisms for public participation in the policymaking process. To be clear, that was not their primary goal at that time.

I founded the Science Cheerleaders, a group of more than 300 current and former professional cheerleaders from the NFL, NBA, and other sports leagues who are pursuing science and engineering careers. From personal experience, of course, I knew that there were a large number of sympathetic minds in this group. I also knew that they offered a terrific opportunity to overturn stereotypical perceptions about the exclusivity of the scientific world. With the support of professional sports leagues, media partners like NBC Sports, the National Science Foundation, Pop Warner youth leagues, and scientific stars like *Why Science?* author Dr. James Trefil, the Science Cheerleaders have become the "superheroes of science" both on- and offline. At the same time, the cheerleaders inspire everyday citizens to connect with science — especially young women who may be considering STEM careers — and work to empower people to weigh in on important science policy discussions.

I created a portal on the Science Cheerleaders website for projects with which citizen scientists could become involved. The combination of the cheerleaders sparking excitement about science with a set of projects that were open to enthusiastic citizens would, I thought, create a process to unite the citizen's desire to be heard and valued, the scientist's growing interest in the public's involvement, and government's need to garner public support. Eventually I imagined these inspired citizens getting more involved in policy conversations and expressing their values and knowledge in influential ways.

To bring more attention to Science Cheerleaders and the citizen science portal, I wanted to mix up the kinds of projects that people could participate in on the site and expand beyond what is traditionally thought of as citizen science. Not that counting birds or bees or monitoring water quality weren't important — far from it. But I wanted to demonstrate the field's incredible diversity to professional scientists, policymakers, and most importantly, to everyday citizens who weren't yet sure how to become involved in science. So the site posted projects in fields as varied as archaeology, astronomy, biology, cybersecurity, epidemiology, gaming, geography, geology, programming, and zoology, among others.

When the number of projects we were posting became unmanageable for hosting on the Science Cheerleaders site, I launched SciStarter.com as a platform fully dedicated to discovering, organizing, and participating in citizen science projects. I wanted to make it easy and fun for people to get involved in projects ranging in commitment from one-off events like swabbing for microbes in Project MERCCURI (an extravagant acronym for Microbial Ecology Research Combining Citizen and University Researchers on the International Space Station) — which David Coil uses as an illuminating case study of citizen science in microbiology in Chapter 6 — to long-term coastal monitoring programs. And the site seems to be meeting a need for engaging people in science and technology. With the help of a network of contributors and media, government, and academic partners, the platform currently hosts more than 1,600 projects and events with more than 50,000 citizen scientist participants and more joining all the time.

Yet even as SciStarter rapidly grew and matured, there still remained the problem of getting the public's voice to be better included in policymaking. I was intrigued by the advances other countries like Denmark and the United Kingdom had made on this front, including inaugurating methods of citizen participation and stakeholder engagement in assessing emerging technologies and science-related issues like climate change. Was something like that possible in the United States?

To answer these questions, I joined forces with Arizona State University's Consortium for Science, Policy & Outcomes, the Woodrow Wilson Center for Scholars, the Museum of Science Boston, and the Loka Institute to found the Expert & Citizen Assessment of Science & Technology (ECAST) Network in 2010. As a collaborative endeavour between academia, informal science educators, and policy partners, ECAST has been instrumental in bringing citizens and experts together to inform and improve decision making on science and technology issues. Our most recent success was in hosting a forum on NASA's Asteroid Initiative, which provided NASA administrators with public perceptions, aspirations, and concerns about the agency's space mission through dialog with a diverse group of informed citizens. Other federal agencies, including the National Oceanic and Atmospheric Administration and the Department of Energy have since enlisted ECAST and together, with SciStarter, we are forging new opportunities for people to move between citizen science and "citizen science policy." Mahmud Farooque's logic model in the final chapter illustrates this vision.

My aim in all this—in the creation of Science Cheerleaders, SciStarter, ECAST, and this book—is ultimately to empower ordinary people to contribute to science, and for their voices to be influential in ongoing science policy debates. It is to cast a wide net through the Science Cheerleaders, to provide opportunities to actually *do* science through SciStarter, and to move people to contribute to related science policy discussions and shape science through ECAST. Citizen science projects give people confidence in their involvement in science, so it's vital that projects connect with people's diverse interests and values in ways that can lead to more profound engagement. This is especially true when citizens seek to change the status quo—scientific, social, or otherwise—as in the powerful social movement-based citizen science that Gwen Ottinger describes in Chapter 5.

There is already broad agreement that our educational priorities for our children must include a greater emphasis on STEM subjects, and I naturally fully support all efforts to encourage this. But I and other citizen science advocates—and many professional scientists—think that concerns about scientific literacy and the influence of public values on science policymaking should be the start of the conversation, rather than the end. There is increasing opportunity today for scientists and policymakers to inform a curious public about the work that they do, rather than assume few would be interested in it or capable of understanding it. But convincing the scientific community and policymakers that the public should be invited to participate in research and decision-making activities is only part of the equation. Convincing the general public—those without an obvious, immediate stake in the outcome of the policy decision—to get involved is still a substantial challenge. Yet I believe that change is coming.

This is not simply the pie-in-the-sky hope of an enthusiastic science cheerleader. Those shad fishermen who worried about declining fish stocks in the Delaware and Schuylkill rivers could see the impact that pollution, overfishing, and dam construction was having on their livelihoods. But more importantly, they could measure this impact by counting the ever-smaller number of fish that were moving upstream to spawn. These weren't just people with hunches; they were citizen scientists with data.

By communicating these observations to policymakers, the shad fishermen provided evidence to support the passage of the Clean Water Act in 1972. Polluting industries were forced to clean up their acts, and fish-blocking dams were altered or removed. Fishery managers placed restrictions on the shad catch, and hatchery operations have released millions of young shad into the rivers. Citizen science-influenced policy helped achieve changes that reflect society's shared priorities and values. The shad that once played such a foundational role in both Philadelphia's ecosystem and economy are slowly returning.

That's the kind of profound change I know citizen science can incite.

Further Reading

Busch, A., & Kaspari, D. C. (2013). *The Incidental Steward: Reflections on Citizen Science*. New Haven, CT: Yale University Press.

Citizen Science Association: http://citizenscienceassociation.org

Citizen Science: Theory and Practice, the journal of the Citizen Science Association: http://theoryandpractice. citizenscienceassociation.org

Cooper, C. B. (forthcoming). *Citizen Science: How Curious People are Changing the Face of Discovery*. New York, NY: Overlook Press

Russell, S. A. (2014). *Diary of a Citizen Scientist: Chasing Tiger Beetles and Other New Ways of Engaging the World*. Corvallis, OR: Oregon State University Press.

PART ONE

1

WHEN CITIZEN SCIENCE MEETS SCIENCE POLICY

Eric B. Kennedy

Read a scientific magazine, visit a university campus, or walk the halls of Congress and you may well hear reference to "citizen science." It's the subject of much discussion, of expanding use in research communities, and even of new federal legislation. According to many, citizen science—put simply, public engagement in scientific research and decision making—represents a radically new way forward: a path that engages every kind of person in research and decision making, democratizes science for all, and offers a new distribution of power and influence in universities and beyond. The term conjures visions of a more inclusive world of science, a more engaged public and citizenry, and rich treasure troves of data for addressing important problems.

Things haven't always been this way. A few decades ago, decisions were largely made by elites—academics, governments, experts—who were insulated from public input or even scrutiny. This rang especially true through the first half of the 20th century, as much of the scientific establishment became increasingly centralized, institutionalized, and separated from the average citizen. Such a shift away from public access occurred for a multitude of reasons. In some cases, like agriculture or food systems, the kinds of invention and innovation required to eke out ever-larger incremental benefits required increasingly complex and well-funded laboratories and research collaborations. In others, like energy production, the technologies were becoming larger and more complex (such as the shift from unique heating in each home to a much larger, more centralized power grid). Perhaps most evidently, the rise of the American military-industrial complex, centered on the atomic bomb, reinforced a concentration of political power among a small subset of scientists and military decision makers.

Even at the time, fears were rising about these shifts. In 1961, President Eisenhower famously warned of the risk that "public policy could itself become the captive of a scientific-technological elite." Eisenhower feared, among other concerns, that the military-industrial complex could divert policy from national security ambitions toward profit motives. Still others, like physicist Ralph Lapp, resonated with fears expressed much earlier. Like President Wilson, who was quoted as fearing "a government of experts," Lapp took exception to the increasing industrialization and centralization of science. In his 1965 book entitled *The New Priesthood: The Scientific Elite and the Uses of Power*, he raised concerns about an emerging concentration of science for ends of military and political power. Although this skepticism wasn't without antecedents, it grew in importance and awareness through the 1960s and laid the groundwork for the citizen science movement that would follow.

The rising credibility of citizen scientists, who displayed mighty influence across a range of social and political arenas in the late 20th century, is different. Where the technology critics expressed fear from afar about the system as a whole, citizen scientists embrace and engage with the system directly. Collectively, they are a group of advocates that seek a voice at the table of power, rather than just knocking over the table entirely. If this is the defining characteristic of citizen science, then the movement is important, powerful, and perhaps even world changing.

Is there enough substance to back up the enthusiasm surrounding citizen science? What roles can citizens, their interests, and their data play in decision-making and governance? And, what does the future hold for this emerging movement? This chapter aims to provide a primer to the subject: free of buzzwords, boosterism, and overly optimistic claims, but grounded in the significant advances that citizen science has made over the short and long terms, and its substantive potential for influencing policy, government, and society writ large. To answer these questions, we begin by considering just what is meant by citizen science, and offering a quick sketch of where this movement sits in the long arc of scientific history. We then turn to the challenges it presents to science, and a consideration of how the citizen science movement might shape policy, how policy may wish to shape the practice of citizen science, and how the two ultimately represent a difficult and important clash.

What Is Citizen Science?

Given the proliferation of citizen science projects, it is easy to think of the movement as having a long history. As citizen science advocates point out, there is a long tradition of regular citizens participating in activities like bird watching (which, on occasion, informed local scientific societies) that, in hindsight, appear as early forms of citizen science. Moreover, there are countless examples of citizens pushing back against elite and expert judgments, from fighting for social causes (like to gain the right to vote or to end slavery) to sociopolitical topics (like countering wars or calling for expanded human rights).

What, then, differentiates today's citizen science from forms of citizen resistance more generally? As hinted at earlier, the mid-20th century represented a time of major shift in the relationship between science and society. In the most profound example, the emergence of nuclear weapons, scientists had effectively attained the power to destroy the world through technology. Moreover, citizens had remarkably limited access to such powerful processes and people. As the scientist and philosopher Michael Polanyi argued in his 1962 article "The Republic of Science," the scientific enterprise had become a real and distinct society, effectively cordoned off from the public. For Polanyi, this implied a requirement to govern the republic in a way that brought about social and scientific good, rather than an establishment entangled with the interests of political leaders. Likewise, scholars like Don Price concurrently (1965) worried that science had become so powerful that it threatened democratic processes and key checks and balances required to protect public freedom. This unquestioned priesthood of science rendered the public voice unwelcome and unnecessary.

In contrast with the aims and methods of other citizen resistance of the mid-20th century, therefore, the citizen science movement took on a particular strategic goal. Instead of tearing down the institutions of power, early citizen scientists aimed to be included and recognized as legitimate experts. This was a move of expansion, inclusion, and, at times, substitution, by inserting and recognizing their new voices over existing power interests. Employing rhetoric, symbolism, and pragmatic strategy, citizen scientists fought to be included in decision making as legitimate partners.

Viewed in this wide lens, citizen science is a remarkably encompassing term. It includes the kinds of projects that are often cited in magazines and the news, such as members of the public counting birds, measuring snowfall, or classifying pictures of galaxies on their computers. It also includes broader efforts towards encouraging public participation in science, whether by hacking or building new devices; partnering with scientists to develop and design entire research projects; or deciding how results of scientific research ought to be taken up in policy. And, importantly, it has blurry edges, wherein many forms of political advocacy — such as members of the public educating other citizens about scientific findings and opportunities for participation, pushing for particular regulatory changes on the basis of scientific arguments, or confronting the actions of companies that may be causing harm — can be included in the realm of citizen science.

Many contest exactly what should be included under the label of citizen science. While some take a broad view, others restrict it more narrowly to particular kinds of activities or contrast it with related areas (like crowdsourcing, prizes, or DIY). Still other voices use their definitions to include, in hindsight, a set of historical antecedents that look alike. To understand these contrasting definitions, it is worth briefly considering the history of how science developed as it did.

How Did Citizen Science Emerge?

The early history of science and technology is largely one of simultaneous co-discovery. Hundreds of years ago, major innovations in how humans understood the world or the tools they invented to interact with it were likely concurrently discovered in many different locations, independent of one another. From a modern vantage, many of these tales of innovation are told as the "great men" stories, highlighting particular individuals, rarely with formal scientific training, who made discoveries or shaped technologies. Yet the historical studies of technologies ranging from bicycles to agricultural breeding indicate much more fractured, complex, and context-dependent stories of technical development. Behind almost all of these mythologized stories of venerated inventors lays a reality where many so-called "average citizens" — farmers, doctors, politicians, and clergy — used innovation to tackled the problems they faced on a day-to-day basis.

One inclination is to look back at either these "great men" or diffuse innovation stories and label them as early examples of citizen science in practice. Yet, although finding such historical parallels may be convenient for a modern movement developing its identity and lineage, neither example seems to capture the essence of citizen science. "Great men" like Charles Darwin or Benjamin Franklin received significant support from the scientific and elite establishments of their time, and leveraged more power and privilege in their scientific passions than the kinds of publics engaged in citizen science today. Nor do diffuse examples of co-discovery of technologies represent citizen science in the modern sense, as the innovation and practice wasn't tied to a desire to participate in scientific communities or government decision making. Early innovators in agriculture, for instance, didn't refine crop-breeding techniques to try to shape government policy, but rather simply to improve their successes at a local, individual level.

Over time, the institutional apparatus of the scientific enterprise grew and formalized. Research became necessarily more complex as goals shifted from initial discoveries in the field towards refining and increasing benefits. This complexity translated into increasingly large research teams, the importance of academic infrastructure to share knowledge (e.g., journals, conferences, and networks), and much more complicated laboratories and equipment to conduct the required manipulations. These shifts had social consequences, including the growth of a much more contained, self-reliant, and defined set of scientific communities, ranging from formal societies to informal notions of who counted as an expert or scientist.

In the American context, it is difficult to overstate the importance of the military — and especially the transformative nature of nuclear weapons as a centralized, scientific project — on these processes of formalization and community definition. Many research projects relied on technologies that were developed in military labs, or were funded via connections to potential military applications. Military culture brought norms of hierarchy and formalization as well, shaping the practices of some scientific fields. The link between science, military, and economic progress resulted in an increasingly tight connection between some forms of science and government decision makers as well. The creation of government agencies and departments responsible for different fields of science (e.g., the Environmental Protection Agency, NASA, or the Department of Energy) further imbued particular forms of science with socio-political power.

This is not to suggest that citizens ever disappeared from the picture entirely. Members of the public were always engaged in various kinds of scientific activities. Many hobbies, ranging from birding to operating ham radios, persisted or grew during this time. Some of them even connected with government or civil society applications, such as participating in the Audubon Society or assisting in times of accidents or disasters.

A particularly vivid example of this kind of public engagement was the "Baby Tooth" study run in the 1950s and '60s by what would become the Greater St. Louis Citizen's Committee for Nuclear Information. Over more than a decade of collection, roughly 320,000 baby teeth were collected to analyze the impact of atomic testing on radiation levels within American children's teeth. The data, which demonstrated that children born in the height of the Cold War had roughly 50 times more Strontium 90 in their teeth than children born 15 years earlier, went on to be used by physicists and physicians alike. It would ultimately contribute to pressure for nuclear test ban treaties.

These efforts had a different character than today's citizen science. For many of the scientific hobbies like birding or ham radio, the urgency for a seat at the table, to be represented, and to be welcomed by decision makers wasn't the defining passion. In the case of the baby tooth study, it was an effort largely organized by community members who already had a background in physics or medicine. By contrast, when people refer to the citizen science movement of today, they are often describing the involvement of otherwise untrained members of the public in policy-relevant questions.

When and where exactly this changed is the subject of many histories (the next chapter by Cooper & Lewenstein, for instance, provides a more thorough account). The broad strokes, however, are clear. By the 1990s, new groups represented the face of public engagement with science. Among the most famous, influential, and now archetypal of this movement were patient advocates in the AIDS crisis. Concerned with how medical research was being done (namely, that many patients involved in clinical trials for new drugs were actually being given placebos — a death sentence), a force of activists emerged. These groups used various scientific positions to argue for new methods that would allow all patients to be given treatment drugs, while still conducting medical research to create new treatments.

As sociologist Stephen Epstein illustrated in his 1996 book, *Impure Science: AIDS, Activism, and the Politics of Knowledge*, these groups took on a new approach to interacting with science. Patient activists didn't take on an interest in medicine as a hobby, nor as grassroots innovation to solve their own problems (like farmers may have experimented with new techniques hundreds of years earlier). Rather, the activists had to lobby a much larger scientific establishment — complete with its giant laboratories, huge funding requirements, drug approval agencies and laws, and massive body of research — to listen to new voices and to take a new direction. This required something different than just casual engagement with science: it required a seat at the table, a voice before decision makers, and serious engagement with elites who had long been separated from the public.

This reveals a key theme in the citizen science movement: broadening judgments of expertise. As much as the citizen science movement was reacting to the growth of the technocratic elite, it largely wasn't for the purpose of tearing down existing experts. Rather, citizen scientists were making a particular claim: that they had a kind of expertise that not only needed to be respected and included in decision making, but that was unique and even inaccessible to existing experts.

At first glance, this sounds like an extreme claim. After all, scientists and researchers are conventionally seen as the real experts; the ones able to discover and share scientific truth with the population writ large. But, for citizen scientists, the truth offered by traditional science was only a small slice of reality. For citizen scientists, the realities that mattered were visceral and close: the industrial plants causing health impacts to the air they breathed on a daily basis, or the doctors who had never known firsthand what it meant to suffer from AIDS. Giving voice to these concerns didn't require tearing down existing experts. Instead, the strategy of citizen scientists was to embrace the value and importance of existing experts (by reading their research, having conversations with them, and becoming fluent in the topics) in a way that built the productive relationships that would allow citizen scientists to gain respect too.

This isn't to suggest that all citizen science projects look exactly like the forms described above. Some examples lean more heavily on activism and particular public-interest agendas (like air quality monitoring or patient activism), while other projects fit into existing scientific projects (like monitoring particular ecosystems or sorting telescopic images of stars). The boundaries between the related concepts of citizen science, crowdsourcing, and activism are increasingly blurry.

Even with these caveats, however, it is essential to understand this tension between the long lineage and recent innovation of the citizen science movement. Citizen science isn't a new phenomenon, as humans have always been engaged in these kinds of sociopolitical struggles over how to innovate in the face of challenges, who ought to make decisions, and how to advance one's own interests. But neither is citizen science a consistent thread through history: the forms of citizen science we see and talk about today are distinctively different in their ambitions of sitting at the table of decision makers.

This particular form of citizen participation has significant power to affect (and be affected by) the political landscape. Sometimes citizen science results appear in the decision-making world as a source of data: a way to gather intelligence and insights about an issue (often, but not always, environmental) in ways poorly suited to traditional scientific study. Other times, the policy is acting on behalf of citizen science, such as legislation currently under preparation and consideration at the U.S. federal level to enable agencies to better incorporate citizen scientists into their activities. In still other examples, the idea of citizen science becomes a proxy conversation or movement for something much more profound: the questioning of how public-government relations ought to work, of what roles citizens and non-experts should be able to play in decision making, and of challenging longstanding norms that have excluded regular citizens from being involved in scientific and technical decision making. In this chapter, we'll survey the field on all three of these issues.

Citizen Science for Policy

As was hinted at in the brief history above, the citizen science movement has often been driven by a relentless passion: protecting a particular species that may be in decline, speaking up for a community of people affected by a significant disease, or monitoring the health of an ecosystem under attack. As detailed in the chapter by Cooper & Lewenstein, these kinds of very specific, local, and grounded concerns initially gave rise to a new voice in science: science driven by community members themselves, rather than outside researchers. In turn, these kinds of projects very naturally led to policy outcomes. Birdwatchers tracking specific species could be leveraged for protecting migratory pathways or breeding grounds. Patient activism could be aimed at reforming clinical trials (like the case of AIDS activists Cooper & Lewenstein discuss). Environmental monitoring could underpin pressure for environmental justice in the face of expanding industry or natural resource extraction.

Citizen scientists, therefore, already have important influence on the world around us through the fruits of their labor. The data collected by citizen scientists has impacted government policy, created new norms and abilities in fields like environmental management, and has even shifted the way that decision makers view the role of the public. These general statements, however, give rise to several specific questions about the nature of the citizen science movement's impact on policy. Where and when do citizen scientists appear in policy? Are the data they produce reliable and sound for making real-world decisions? How might this change as citizen science grows as a movement? And are there possible harms or challenges arising from integrating this kind of research?

To answer these questions, we must begin with a quick introduction to the variety of projects and initiatives that may fall under the banner of citizen science. Even the most cursory introduction reveals something of a schism within citizen science as a whole: what exactly counts as an example of a citizen science project, versus any number of related concepts like citizen activism, hackers, DIYers, or hobbyists.

For some, citizen science is defined through a variety of narrower lenses. This understanding of citizen science largely draws on historical "quintessential examples," often grounded in the world of animal and environmental monitoring. This view of citizen science leans heavily on data collection and, to a lesser degree, data processing. The citizen scientist serves as an educated, volunteer researcher, albeit limited in his or her scope: collecting various kinds of readings, counting different animals, or perhaps performing somewhat rudimentary kinds of analysis—especially those that are straightforward for a human, but difficult to automate or massive in scope (e.g., sorting through hundreds of pictures to identify particular kinds of images).

A characteristic example of this kind of project is wildlife monitoring. The Monarch Joint Venture, for instance, links together more than a dozen different regional projects that monitor monarch butterflies, their habitat, and their migratory patterns. Citizen participants can join one of these projects to record monarch sightings, share photographs, or even identify monarch eggs. The data are often collected in conjunction with a university or nongovernmental organization, allowing the observations of a large number of individuals to be aggregated into larger patterns.

As the movement has grown over the past decade, however, a number of its proponents have adopted a much more encompassing approach. As a community, these citizen scientists and organizers have taken on a "big tent" mentality, characteristic of a movement more eager to grow its numbers, allies, and momentum than to prioritize demarcating what does and doesn't count as an example of citizen science. This broadening embraces a wide variety of commitment levels (from participating in a one-time bird count to tracking and sharing personal health data for a lifetime), project scopes (from recording plant counts via pencil and paper to launching cube satellites with various instruments), topics (from the microbes between your toes to sorting telescopic images of far-away galaxies), levels of involvement (from collecting data for existing ecological research projects to co-organizing and developing a community-led project to address concerns about pollution in your neighborhood), and activities (including do-it-yourselfers, hackers, and activists).

This broadening also means that citizen science projects can have a wide array of strategic purposes. Conversations with the Environmental Protection Agency (EPA), for instance, illustrated at least four roles in which they've seen successful citizen science projects:

- **Empowering communities**, by encouraging citizens to take an active role in collecting, processing, analyzing, and applying information—and encouraging new groups to participate and engage, especially those who have previously been marginalized or excluded.

- **Establishing ongoing monitoring**, especially where citizens are able to collect (or have already started collecting) much larger, more detailed, more thorough, and more regionally appropriate data than the agency could collect on its own.

• **Extending research** to questions, areas, and topics that used to be beyond the capabilities of a government agency (e.g., new species, locations, or questions), including using citizen science to solicit various kinds of public input (such as blended research and deliberation activities).

• **Educating** citizens about environmental (and other) issues, through first-hand experiences that teach participants about both science generally and the particular topic of study.

Even this range belies the full variance among citizen science projects that have proved successful. Some citizen science projects, for instance, involve citizens in later stages of the work, such as interpreting and visualizing results according to their values, perspectives, and priorities. Still others integrate existing datasets (like satellite imagery) with citizen contributions (such as having the public 'adopt a pixel' of the satellite image, and submit a panorama from that location to provide additional data). In short, citizen science efforts have expanded to cover a tremendous range of research. Yet, this expansion gives rise to several questions about the potential flaws and limitations of using citizen science to inform public policy and decision-making.

Can We Trust Citizen Science?

One perennial question of citizen science is whether its data can be trusted. Is the information gathered by citizens—especially if they're relatively untrained compared to the scientists in the area, or if they have particular activist interests—reliable enough to be the basis of policy making? Yet, this question must co-exist with a larger quandary: just how much can we trust science in general?

In conversations with various government agencies and non-governmental organizations, questions about whether citizen science results can be trusted were largely met with rebuff and frustration. The generalized reaction was straightforward: real challenges with the data quality of citizen science parallel the challenges faced in all scientific work. In other words, the results produced by citizen scientists are not alone in needing to be carefully scrutinized on the basis of their methodology, quality assurance, context, and application. Like all scientific findings, they are best served by careful evaluation on a case-by-case basis, and by integrating many different sources of data.

Questions surrounding trust in science co-evolved with the rise of the technical elites described earlier this chapter. One reason was the increasing complexity of the technoscientific challenges at hand. While some scientific questions had very clear answers that were easy to validate, many of the problems of the 20th and 21st century were much more complex. In particular, the fields of health and social sciences offer particularly vexing problems. Finding the cure for diseases that are a combination of genetic and environmental factors, for instance, is difficult to solve quickly or conclusively.

Another reason is that the very establishment of a technoscientific elite excluded many from participating in science. Medical studies largely tested cures on male subjects over females. Many of the research questions considered tended to be those of interest to well-to-do populations over the impoverished (e.g., vastly more money goes into cosmetic R&D in America than into many deadly diseases in sub-Saharan Africa). Decades of exclusion have not only led to a distrust of the scientific enterprise, but also serious concerns over the legitimacy and accuracy of work that focused on an overly-narrow population.

In short, the question of whether we can trust science is just as difficult—or more difficult—as whether we can trust citizen scientists. Moreover, there is reason to believe that including citizens in the scientific process may well help address both of these concerns. Expanding the scale and scope of research through citizen volunteers may well ameliorate many complex challenges. Furthermore, if citizens are involved in the process of selecting and defining research problems, topics that had previously been excluded from consideration by the scientific enterprise may be investigated from important new perspectives.

Beyond this "skittishness" around the use of citizen science, other open questions remain about the potential of integrating citizen science into policy. As with all scientific data, a major challenge is translating findings to policy makers in a way that actually influences decision making. This is compounded by the need for additional expertise in evaluating and integrating citizen science data into other data sources, especially for researchers with little familiarity with these emerging techniques. Citizen science projects must also be relevant and timely, which can prove challenging given the long timescale of many such projects (especially those focused on long-term monitoring, but also in general, given the need for training participants, allowing time for the research to be conducted, and synthesizing results). These kinds of challenges also help to separate the particular areas where citizen science is the ideal approach from those where it may be poorly suited.

Data "Fit For Purpose"

While the spread of citizen science is certainly interesting, perhaps most important is the creation of techniques "fit for purpose," a phrase introduced to me by researchers at the EPA. Under a fit for purpose model, citizen science is neither inherently good nor bad. Moreover, citizen science should not be blindly adopted with the refrain assigned to emerging computational power in the 1990s: "better, faster, cheaper." What matters instead is finding the best method for answering a given research question or achieving a particular aim, whether that method is citizen science or something else. By focusing on finding the best tool for a given task, the relative advantages of each approach can be maximized.

For some purposes, for instance, very carefully planned representative samples are important. Forming environmental or health regulations, for instance, requires kinds of knowledge generation that may often be best suited to well-equipped labs or large-scale, centrally organized trials. By contrast, citizen science is often the most appropriate method of identifying potential problems, flagging areas for further research, establishing long-standing baselines, or meeting the goals of empowerment, extension, and education. In many cases, networks of citizen scientists may already have established extensive datasets, professional skills, or lived experience that help improve the robustness and timeliness of future work. In other situations, like backyard monitoring of snowfall levels, citizens may be able to offer data at a much larger scale or thorough resolution than traditional measures. Some problems may incentivize particularly accurate and committed public participation, such as homeowners monitoring the groundwater quality in their wells, which can benefit both their own family and environmental agencies.

Largely, however, efforts to include citizen science in policy have been relatively bottom-up. As opposed to top-down directives to add citizen science approaches to existing projects, most of the spread of citizen science has been driven by passionate researchers and participants creating and expanding projects. There is also variation between the approaches of different government agencies because of their differing data needs, agency missions, and research capacities. While this has led to perhaps a slightly slower expansion of total citizen science initiatives relative to the desires of activists, it is largely advantageous: case-by-case integration is closer to a "fit for purpose" approach, where citizen science is integrated when it is most methodologically appropriate to the task at hand, and where small-scale experimentation can help researchers learn what's possible, improve on their methods, and develop new approaches to meaningfully include citizens.

Unintended Impacts

One concern with the rise of citizen science is a fear that it might undermine or replace government-run monitoring programs. In this scenario, it becomes more difficult to justify the establishment or continuation of an expensive research program (and the richness of professional knowledge, context, and capability with it) when the data can be generated by free labor from volunteers. These concerns also highlight broader questions about the potentially exploitative nature of citizen science, particularly if its participants are subject to personal liability while conducting the work, or if their efforts are broadly shared without compensation.

Several of those organizing citizen science projects, however, suggested that it is far from likely that citizen science efforts are sufficient to displace government-supported initiatives. Even if citizen science projects increased in scope by orders of magnitude, they are largely complementary with existing research work. Considering a broader definition of citizen science as well, such as the development of lower-cost monitoring tools, citizen science offers the potential for the expansion of existing data generation projects, enabling government investment to achieve a higher dollar-for-dollar value. Moreover, they sustain a significant amount of synthesis and administrative work to develop, maintain, and share the work of the projects, as well as integrate it in the larger landscape of science and science policy.

A more pressing question for many, however, was the implication of citizen science on the policy process itself. If citizen science's growth—and the resulting proliferation of cheap sensors, community activism, and public participation—is as successful as hoped, it challenges today's models of science policy. These tools enable a very direct and quick-paced avenue of interaction with governments, especially when compared with the relatively slower pace of traditional academic research on topics like health and environmental safety. In many cases, the data brought by activists may be non-interoperable with existing data, or represent a very different kind of knowledge than long-term monitoring citizen science projects: a one-time spike leveraged as a call for urgent action, for instance, versus a rolling average. Such examples will inevitably force a rethinking of many facets of scientific decision-making—fueled by citizen science, and similar kinds of community participation.

Citizen science is also symbolic of a broader shift in contemporary research toward "big data" and the use of tremendously large datasets. Citizen science—particularly those projects involving automated sensors or large-scale participation—has the potential to yield huge volumes of data. This pipeline can prove challenging, both with respect to cleaning, integrating, and synthesizing data, but also in terms of data storage, retention, and long-term use. Like all data-heavy efforts, the use of large-scale citizen science data will require the rethinking of many logistical questions about research, and retraining and retooling of parts of the research enterprise.

What is clear, however, is that simply denying the emergence of citizen science and these challenges is untenable. The movement is growing, and will continue to challenge governments and institutions alike. It is essential, therefore, to begin anticipating the impacts of citizen science, and the potential policy options for developing it to meet society's needs.

Policy for Citizen Science

It is common to frame the intersection of citizen science and public policy as a straightforward question: What can the world of policy and governance do to support the emergence of citizen science? This problem definition at best offers a simplistic representation of the issues in play and, at worst, a misrepresentation that obscures important areas of open contention between the two worlds. To view citizen science as an unadulterated good obscures the rich and complex tensions that arise as citizen science's purview — and corresponding influence on policy — expands. Moreover, such a view risks replicating many of the very problems citizen science seeks to address: a monolithic view of what counts as "good" scientific practice, the institutionalization of particular voices in particular roles (and the corresponding exclusion of others), and a precommitment (at times, independent of context) to particular methods of scientific practice.

This is not to suggest that the government does not have a vital role to play in empowering citizen science efforts — nor that these tasks are unimportant. Many challenges with integrating citizen science into public policy are already known, at least to a degree, and can be addressed through various government policies. In conversations with experts and thought leaders in citizen science, many emphasized a similar series of questions and concerns. For many agencies and employees, one of the most direct limiting factors is time. Overworked, under-resourced, and under significant pressure to fulfill a number of objectives, agencies have little time to invest in learning new methods and developing new procedures to integrate a new kind of scientific practice. This naturally leads to a more diffuse, decentralized approach, wherein agencies that are supportive tend to encourage the inclusion of citizen science on a case-by-case basis, where researchers are willing and able to do so through their own efforts. While this could theoretically lead to strong, contextually appropriate applications of citizen science, it also can undermine the need for systemic integration, mutual learning, and the implementation of citizen science in the places where it might be most beneficial (rather than in the places where the project leaders are the most interested).

This pressure on time and resources is augmented by a frequent lack of institutional support. Until recently (and even now, in many agencies), those individuals interested in pursuing citizen science have largely had to develop their skills on an individual basis. Moreover, because citizen science is often not the norm, they are left pioneering integrative methods on their own, including addressing supervisor hesitation, dealing with questions about data integration and access, and having to defend both their own projects and citizen science generally.

A plethora of ongoing and emerging efforts support the development of citizen science and address these kinds of challenges. Like on other policy issues, these tools include signaling support in formal and informal ways, providing resources for the expanding community, and providing clarity in laws and precedents.

The first step towards developing pro-citizen science policy is through formal and informal methods of signaling. Some of these efforts are large-scale, such as the *Crowdsourcing and Citizen Science Act*, a bill which was introduced in the United States Congress in 2015. While the act itself offered a degree of formal support, because it was relatively thin on specific deliverables (e.g., funding or requirements), its informal role was perhaps more significant: to signal to government agencies that citizen science was something to encourage, pursue, and incorporate into their formal procedures. Indeed, much of the signaling and formal implementation rests with agencies themselves, many of whom (including agencies like NASA and the EPA) have begun efforts to include citizen science in some projects, to participate in networks, workshops, or conversations about citizen science, or simply to allow their staff to comfortably consider citizen science on a case-by-case basis.

Once a government has signaled openness toward a particular kind of activity, more practical questions of resourcing can come to the fore. In this case, efforts at resourcing citizen science in the U.S. government have largely leaned on experience-, advice-, and resource-sharing initiatives, as opposed to large-scale financial investment. Experts at the White House and several agencies, for instance, have been eager to share the value and successes of "Communities of Practice" established to allow their employees to share best practices, lessons learned, and methods and models for replication. Several other initiatives, both domestically and abroad, have worked to develop "citizen science toolkits" to enable government agencies and researchers to more smoothly integrate citizen science approaches into new and existing projects. Indeed, such toolkit efforts have become so prominent that one expert in the field recently expressed frustration that yet another group was "going to spend two years rebuilding all the same toolkits" to support beginners at citizen science projects, when the expert would rather "see people roll their eyes when you talk about why [citizen science] is an amazing and awesome thing." Some financial support may become available as well, especially as the world of funded prizes (e.g., large prizes designed to incentivize private-sector solutions to vexing challenges, such as private space flight through the X-Prize) begins to both interact with and integrate with citizen science initiatives.[*]

The third role the government can play to support citizen science is by establishing clarity in precedents and in laws in such initiatives. Several agencies, for instance, identified internal concerns about the potential legal vulnerability that could arise from citizen science projects for a host of reasons: issues like intellectual property, liability for volunteers, data sharing, and legislative and regulatory impact. Although large-scale legislation like the *Crowdsourcing and Citizen Science Act* can play an initial role in signaling the acceptability of citizen science, the actual battles of liability, open-access, and unexpected impacts will likely be fought in the weeds—at the level of local cases, specific implementations, and one-off challenges. Indeed, it is these questions that begin to hint at the more complex nature of the citizen science and policy interface.

Alongside many of these logistical challenges with citizen science—and the government policies and initiatives designed to address them—there remain much deeper tensions between citizen science and the policy environ-

---

[*] Prizes also offer an interesting case of the power of government signaling. An American government official reported that one act (America COMPETES) authorizing the use of prizes led to an order of magnitude increase in the number of prizes being offered. Yet, the official noted, only about a quarter of those prizes officially cited the authorization of the act. Three quarters of the new rush resulted, at least in part, from the informal signaling of the openness to and ability of prizes, rather than the formal authorization to use them.

ment. As is explored further in the next section, in some ways citizen science represents a fundamental confrontation with the policy world. It reinvigorates questions about who ought to be involved in making decisions, how they ought to do so, and what norms and expectations define this new world. These subversions lead to challenging tensions that show up in seemingly unrelated policy questions, like those of legality, ethics, and the "ideal" scientific method.

One set of tensions includes the legal dimensions of citizen science. In a traditional policy-making environment, those conducting research or working on behalf of decision makers are often formal employees or contractors. Over the past century, these groups have expended considerable effort to clarify the laws that they operate under, and to ensure that those laws afford protections to both the employers and the employees. When it comes to citizen science, however, uncertainty remains about the legal responsibilities and liabilities associated with the inclusion of volunteers in scientific projects. Both the issues and perspectives are numerous. Who is liable for injuries or costs resulting from accidents involving citizens participating in such projects? How might this change with different levels of formality or agency involvement (e.g., hobbyists reporting data they collect during long-standing hobbies, versus citizens actively recruited as data collectors for a particular government-supported project)? What kinds of measures must be taken to ensure volunteer safety? Who has the responsibility for overseeing these measures?

Such legal questions are fundamentally ethical as well. While many government agencies are rightfully concerned with questions of liability, these concerns are often proxies for underlying ethical or moral struggles about treating other individuals and communities in virtuous, positive ways. These questions become complex in the world of citizen science, where data are often highly personal (e.g., health data collected from wearable technologies), sensitive (e.g., data that, even in aggregate, could provide companies or governments with knowledge about your daily activities and values), or significant (e.g., knowledge that is closely guarded by a community or indigenous group). The use of the data is similarly ambiguous: under which conditions, towards what ends, and over what timespans can the data be investigated?

Although efforts by institutional review boards (IRBs) and progress towards so-called "open data" can help address some of these concerns, they ripple forward into their own challenges. Though the protections offered by IRBs and ethics procedures address some of these questions, they also result in a very different regulatory environment between the United States and other locations — the former's reliance on IRBs and other institutional mechanisms undermining the kind and pace of innovation in citizen science methods that some experts report seeing in Europe. (Take, for instance, the Paperwork Reduction Act, meant to reduce the paperwork burdens placed on individual citizens, which one government official cited as adding approximately a year — of paperwork, ironically — to the process of developing a citizen science project.) Moreover, while open data movements can help to address some concerns about access to one's own data (and are sometimes required by federal law), it also opens the floodgates of potential data use by unintended audiences, such as corporations mining data for commercial purposes.

In short, while some boosters of citizen science may argue that the role of policy is to clear the landscape for the proliferation of such projects, the reality is much more nuanced. Questions of "policy for citizen science" are not simply questions of whether citizen science should be permitted or encouraged. Rather, they're much more open deliberations about the kinds of citizen science that should be encouraged, how it should be managed in particular (and sometimes particularly sensitive) contexts, and the unintended consequences it may have in future applications and unexpected uses. Indeed, this suggests a more fundamental question to be asked of citizen science: how does it challenge our traditional norms of government and policy, and how should we respond to those challenges?

Citizen Science Meets Policy

As has been demonstrated thus far, citizen science not only has tremendous potential for influencing policy, but already plays a major role in many existing avenues of research. Policy can serve as a lever to address the limiting factors, but it ultimately must engage in a more nuanced conversation about the particular forms of citizen science we wish to enact and the values behind them. This conversation is driven in large part because of the different ideals espoused by each field: the expert-driven process of traditional public decision making, and the citizen-driven, non-hierarchical nature of many forms of citizen science.

This challenge by citizen science towards the norms of traditional government decision-making is part of a much larger call for renewed and reinvigorated forms of democracy. This movement is certainly most prevalent in the American context, but is echoed in the Commonwealth, in Europe, and around the globe. It is found in the emerging push to include previously marginalized communities in government decision-making and scientific processes; to increase participation in democratic processes among youth and other communities; to make data and research more openly accessible and freely available; and even to hold governments to account for overreaching programs of surveillance or policing.

In many ways, citizen science captures the best of intentions behind these calls. It pushes for a science that is inclusive, open, and transparent. It calls for the remaking of expert processes in ways that invite new members to the table, and that allow avenues for all communities to feel ownership and opportunity when it comes to scientific processes. And it holds research—especially publically funded research—to account through demands for open access, open data, and open lines of communication with researchers and users. Ultimately, it aspires to open pathways for participation in science, and to influence decision making, to anyone who seeks them.

These efforts challenge not only the norms of the scientific enterprise, but many of the values of government. The efforts of citizen scientists to participate are no longer restricted to data collection. They call for a chance to participate in designing research, determining which questions get asked (and where, and how), and influencing how these data shape decisions and actions far into the future. The significance of these desires is not lost on the citizen science advocates seeking them. On multiple occasions in conversation, for instance, citizen scientists and allies described the "subversive" power of citizen science to "catalyze a revolution" such that "participatory methods become the norm." These are pressures that will fundamentally reshape science, policy, and government as we know them.

Exactly how these institutions will be reshaped is much more difficult to say. At this stage, it is clear that our institutions are struggling to respond to these new forms of data and process. Some of these struggles are fairly predictable in the context of an emerging movement, like many long-established researchers who may question the validity or value of citizen science in their work. Many struggles are much more substantive and offer pause to even those advocating for the emergence of citizen science. Agencies wonder how new forms of da-ta—that don't comply with well-founded standards for measurement, for instance—ought to be integrated into the processes of creating, changing, and enforcing legisla-tion. Others are concerned with the rapid pace of citizen science, which often is alien to the moderate tempos of many scientists, labs, and long-term studies. Still more wonder about the representativeness of citizen science, and whether particular citizen science projects risk mar-ginalizing some people from participation in this new wave of efforts.

The very act of surfacing these questions has value. Questions of validity, standards, process, representative-ness, and inclusivity have always been central to the scien-tific enterprise. These are not—or at least should not be—new questions, even if they have long been obscured or overlooked in the quest to "do science." What kind of data should drive regulatory processes, for instance, should not be a taken-for-granted assumption, but should rather be subject to explicit, regular, and reflective consideration. Determining which research questions are worth asking, what kinds of methods should be used to answer them, and which people and communities might have relevant knowledge—these should all be questions scientists ex-plicitly engage with through all stages of their research.

In short, whether one conceives of citizen science in a narrow way, or much more broadly, it is clear that it chal-lenges our norms about how science and governance should be done. In none of these challenges should we assume that citizen science offers a perfect or complete answer, or an alternative model that should be adopted wholesale and without reflection. Rather, it is the process of asking these questions and engaging in explicit reflec-tion and redesign that is most valuable.

Conclusion

The world of citizen science is rapidly emerging and, in many contexts, is already upon us. It's a massively diverse field, ranging from projects that engage the public in bird watching or air quality monitoring, through to research efforts where members of the public themselves drive the very design of the entire project. Citizen science has provided an untold amount of data for research, and has galvanized efforts up to and including a Congressional act to support its growth.

In the chapters that follow, we offer a glimpse of the past, present, and future of citizen science. From a history of the field to several in-depth examples of its use today, this book provides a quick guide to what people mean when they refer to "citizen science," and how it might impact everything from government laws to private businesses, and from worldwide environmental monitoring to your very own health. At the same time, the chapters offer implicit and explicit connections back to the questions and topics raised in the past few pages. How will citizen science affect the decisions made in this world? How might governments and other institutions try to empower or shape it? And how might it cause us to reimagine and redesign government and society itself?

Citizen science is neither intrinsically good nor bad. It has the potential to make a life-and-death difference for communities that have long been excluded from decision making, while also containing the possibility of tremendous risk and challenge. It's an essential topic to know and understand for two reasons. First, citizen science and its related movements are already having a transformational impact on science, often for the better, and are bound to expand in scope. Second, and perhaps most importantly, citizen science has the power to raise many questions about the who, how, when, where, and why behind science and decision making—questions that are long overdue to be asked, and incredibly important to grapple with.

Further Reading

Bowser, A. and Shanley, L. (2013). *New Visions in Citizen Science*. Wilson Center, Case Study Series Volume 3. https://www.wilsoncenter.org/publication/new-visions-citizen-science

Corburn, J. (2005). *Street Science: Community Knowledge and Environmental Health Justice*. Cambridge, MA: MIT Press.

Federal Crowdsourcing and Citizen Science Catalogue: https://ccsinventory.wilsoncenter.org/

Gellman, R. (2015). *Crowdsourcing, Citizen Science, and the Law: Legal Issues Affecting Federal Agencies*. Wilson Center, Policy Series Volume 3. https://www.wilsoncenter.org/publication/crowdsourcing-citizen-science-and-the-law-legal-issues-affecting-federal-agencies

Haklay, M. M. (2015). *Citizen Science and Policy: A European Perspective*. Wilson Center, Case Study Series Volume 4. https://www.wilsoncenter.org/publication/citizen-science-and-policy-european-perspective

Ottinger, G. (2010). "Buckets of resistance: Standards and the effectiveness of citizen science." *Science, Technology & Human values* 35(2): 244-270.

2

TWO MEANINGS OF CITIZEN SCIENCE

Caren B. Cooper and Bruce V. Lewenstein

An Initial Story of Citizen Science: Democratized Citizen Science

In 1981, AIDS was recognized as an epidemic. In 1985, the HIV antibody test became available to the public. Before there were effective treatments, people without symptoms were learning that they were infected. Seasoned activists in the gay community came to realize that the future of their health required a close working relationship with immunologists, virologists, molecular biologists, epidemiologists, and physicians.

These AIDS activists took a four-pronged strategy to gain credibility and authority. First, by attending conferences, critiquing research papers, and receiving tutoring, activists learned the language of researchers and pharmaceutical companies, and the culture of medical science. Once activists were able to talk about viral assays, reverse transcription, cytokine regulation, and epitope mapping, scientists were receptive to discussions. Second, activists represented people with HIV/AIDS and helped ensure that enough people would enroll in treatment trials and comply with protocols to make the trials scientifically useful. Third, activists shifted the discourse away from historic abuses of clinical trials tainted by lack of informed consent, moving instead to a conception of experimental treatments as a social good to which everybody should have equal access. They argued for the right of human subjects to assume the risks of experimental therapies and to be informed partners in research. Finally, activists and researchers who believed drugs should be tested in real-world situations with heterogeneous groups changed the protocols for clinical trials.

Ultimately, research improved as treatment activists— members of the lay public—influenced not only the design, conduct, and interpretation of clinical trials, but also the speed with which they were carried out. On the basic premise that AIDS clinical trials were simultaneously research and medical care, the timeframe for testing the safety and efficacy of AIDS drugs was reduced to months, rather than years (adapted from Epstein, 1995).

A Second Story of Citizen Science: Contributory Citizen Science

eBird is a free, online citizen-science project that began in 2002, within which a global network of bird watchers contribute their bird observations to a central database. Well over three million people have engaged in eBird: in 2015 alone, over 1.5 million people engaged with eBird via the website or mobile devices. Over the years, 270,000 participants have submitted data (<10%) and an estimated 1% have submitted 99% of data. The 1% includes the world's best birders as well as less skilled but highly dedicated backyard bird watchers. Since 2006, eBird has grown 40% every year, which makes it one of the fastest growing biodiversity datasets in existence. It has amassed more than 280 million bird observations from almost two million locations, with observations from every country on the planet.

The most frequent use of the eBird database is through handheld apps that people use to figure out where to go birdwatching. In the early years, 2002-2005, with the slogan "Bird-

ing For a Purpose," the project failed to engage a sufficient number of birders. In 2006, project managers changed their strategy, and introduced the tagline "Birding in the 21st Century." The shift in philosophy, as illustrated by the shift in slogans, made the project successful. eBird moved away from appealing to a birder's sense of duty, succeeding instead by helping birders embrace the excitement of getting better at their hobby while simultaneously impacting the future.

The project leaders showed other birders how using eBird makes them better birders. And better birders make better science, because they provide better data. For example, they submit complete checklists. Initially, 75% of submissions to eBird were incomplete checklists; now over 80% are complete. The last two State of the Birds reports (created by a coalition of conservation organizations) relied on eBird data to examine species occurrence, habitat types, and land ownership at a level of detail never achieved before. These reports inform decisions of the U.S. Fish & Wildlife Service and the U.S. Forest Service. The Nature Conservancy uses eBird data to identify which rice farmers in the Central Valley of California they should ask to flood their fields at just the right time for migrating waterfowl. Researchers have written more than 100 peer-reviewed publications using eBird.

The examples above can both be described as "citizen science." The first fits a use of the term introduced in the mid-1990s by British sociologist Alan Irwin (1995) to describe a more democratic, participatory science. The second fits a use of the term that can be traced back to Rick Bonney (1996), then a program director at the Cornell Laboratory of Ornithology, as he tried to describe projects where nonscientists contributed scientific data. Below we explain these two meanings of the phrase citizen science, give a brief account of how they emerged, and explore the sweet spot where the two overlap.

Alan Irwin: The Citizenship of Science

In the late 20th century, historians and sociologists of science increasingly understood that science is embedded in the fabric of society. Consequently, some aspects of science are shaped by major threads in that fabric. For example, institutional forces (such as military and corporate interests) may dominate scientific agendas, instead of the agendas representing the needs and desires of broader publics. One can see this in the way that interests of the pharmaceutical industry drive much research on cures for cancer, even though some public interest groups suggest that we need more research on the environmental causes of cancer. Irwin's work—in a 1995 book titled *Citizen Science*—addressed the varied social pressures shaping science by seeking to reclaim two dimensions of the relationship of citizens with science:

1. Science should address the needs and concerns of citizens, and seek to meet those needs.

2. The process of producing reliable knowledge could be developed and enacted by citizens themselves. People bring into science such things as local contextual knowledge and real-world geographic, political, and moral constraints generated outside of formal scientific institutions.

Though Irwin's idea of a more democratic science has been widely used by scholars in the sociology and politics of science, his use of the term "citizen science" did not itself acquire scholarly cachet. Instead, researchers came to use terms like "activist science" or "public engagement."

Rick Bonney: Contributing Observations to the Scientific Method

The second meaning of citizen science developed in ornithology, when Bonney used it to describe birdwatchers' voluntary contributions of observations across North America. According to Bonney, the term came to him as he stared out the window while writing a grant proposal in 1994 to support collection of those contributions. He used the phrase publicly in a 1996 magazine article, not knowing about Irwin's work. "Citizen science" became widely used at the Cornell Lab of Ornithology; then, as the Lab of Ornithology developed new projects and connected with analogous volunteer efforts by other organizations, the term spread.

In 2014, the Oxford English Dictionary documented that the phrase citizen science was actually used before Irwin and Bonney. In 1989, the National Audubon Society used the term in a way similar to Bonney's use—in their case to describe a program where volunteers collected rain samples, tested the acidity levels, and sent results to Audubon headquarters. The OED defined citizen science as "the collection and analysis of data relating to the natural world by members of the general public, typically as part of a collaborative project with professional scientists."

Thus the earliest use of the term described projects in which a professional entity designed a scientific project and geographically dispersed volunteers contributed observations, usually in ways that aligned with their hobbies and interests. Because of the large number of these projects, the term has most frequently been equated with these top-down projects, with an emphasis on volunteer data contributions. More recently, the term has been used to describe a wide variety of styles in which the public helps carry out any of the steps of the scientific method, whether conceiving of the research questions, designing methods, collecting the data, and/or interpreting results. These other styles include projects that involve more collaboration between scientists and nonscientists in project design and even projects that emerge from community needs with only advisory input from professional scientists.

The second usage of citizen science gained popularity with the media. By the early 21st century, a community of project developers sought to unite various public engagement practices into a professional field of practice. But in the process, these practitioners recognized drawbacks in the term citizen science. Some felt the word "citizen" excluded those not claiming citizenship in the country where they contributed to projects (such as migrant workers engaged in community-based forestry to sustainably harvest salal, a non-timber forest product used in the floral industry). Others felt the term only pertained to contributory style projects and therefore excluded community-based projects (such as projects monitoring polluted waste emerging from industrial plants). Still others felt the term required the abandonment of other terms with longer histories, such as participatory action research and community-based management.

Although each of these critiques raised issues similar to those addressed in Irwin's 1995 study, few people in the practitioner community knew of that scholarly work. In 2009, an effort was made to identify a broader term, renaming the field as public participation in scientific research, or PPSR (Bonney et al., 2009). But by the middle of the second decade of the 21st century, the term citizen science had become the most popular, with little recognition that the phrase unintentionally co-opted Irwin's original intent.

The History of Citizen Science

Although the term is relatively new, the practice is old. The professionalization of modern scientific practices occurred in the late 19th century. Before scientists were called scientists, they were called men of science and natural philosophers. Before citizen science was called citizen science, the practice of gathering observations by enlisting the help of hundreds, even thousands, of ordinary people was not referred to by any particular term at all. Even relatively recently, when initiated by conservation organizations like the North Carolina Wildlife Resources Commission in 1982, the practice of enlisting the lay public in monitoring beaches and collecting data on turtle nesting was simply termed volunteer monitoring. The practice of what is now frequently called citizen science did not begin with the coining of the term. We can use the term citizen science with historic activities and see that leaders in science in the 18th and 19th century carried out citizen science.

In 1776, Thomas Jefferson made plans for the collection of weather data across the state of Virginia via what would today be called contributory citizen science. Beginning in the late 1840s, U.S. Naval officer Matthew Maury created maps of the seasonal distribution of whales and of ocean wind and currents by aggregating observations reported by thousands of military and merchant vessels. William Whewell, Master of Trinity College, won a Royal medal for work based on almost a million observations of the tides systematically, and synchronously, collected by lay people on both sides of the Atlantic Ocean in 1835. Denison Olmsted, professor at Yale, crowdsourced meteor observations in 1833.

So for more than 200 years, scientists have been crowdsourcing observations. Today citizen science is an umbrella term under which to describe a practice that is occurring in many disciplines in which volunteers collect and/or process data. Many whole fields have long histories of such a practice. For example, the longest-running meteorological records in the United States were collected by volunteers in the National Observers Network, which started in 1880. The longest running ornithological surveys in the United States have been carried out by bird watchers in the Christmas Bird Count since 1900.

Other fields have shorter histories of such practices, with the speed and ubiquity of communication and information technologies assisting in creating new research frontiers. For example, volunteers in exclusively online projects solve three-dimensional puzzles of protein folding, individuals use automated sensors to detect earthquakes, and indigenous people in non-literate communities use smartphones to map important natural resources. Public and scientist collaborations continue to expand using a variety of labels: for example, community-based natural resource management, participatory action research, participatory forestry, and volunteer geographic information. Despite its limitations, citizen science provides a useful, catchall term for all contemporary activities in which the public is involved in the scientific method.

Yet there is still the question of how the "public participation" type of citizen science links with the "democratic action" idea introduced by Irwin. If the term has come to represent a multitude of ways that the public is involved in science, what distinguishes it from Irwin's initial intent? In the practitioner world, citizen science originally referred only to participation in data collection; it then expanded to practices that include the public in other aspects of the scientific method, such as formulating the question and interpreting data. Nonscientists, however, can engage in the production of reliable knowledge (also known as "science") in ways other than contributing data. Publics can pose original questions. They can identify relevant variables and sources of data that professional scientists would miss. They can shape the norms and practices established around the scientific enterprise of validating knowledge. Each of these contributes to a more democratic vision of science embedded in society. While the "participatory" version of citizen science describes how people can serve as instruments in the scientific method, the "democratic" version shows how people can influence and transform the larger scientific enterprise.

The term is coming full circle. Increasingly, practitioners of the "participatory" citizen science see "democratic" citizen science as their goal. Particularly in projects that involve environmental monitoring and environmental justice, practitioners and participants seek to transform the power dynamics of local, regional, national, and even international communities. They seek to exercise power that challenges the interests of large government, corporate, or even academically-based research communities.

The technical details explored in the subsequent chapters of this book largely relate to crowdsourcing observations by the public (the participatory model). But ultimately, a larger reason for refining citizen science methods is to increase capacity for research agendas to align with public interests. Practitioners of citizen science seek a hybrid of the Bonney/OED and Irwin meanings: essentially, a gold standard for citizen science practice in which people do more than contribute data, and researchers do more than use the data. Together, a new relationship between scientists and the public will be created. Citizen science strives for designs that will achieve what Irwin envisioned with his original use of the term: scientists engaging with people in ways that deeply shape what we know about the world.

A Third Story of Citizen Science: Democratized and Contributory

With a phone call to Marc Edwards, an engineering professor at Virginia Tech in April 2015, LeeAnne Walters, a resident of Flint, Michigan, set in motion the development of the Flint Water Study, a citizen science project to measure lead levels in tap water. Walters was a stay-at-home mom who could not get state or local officials to respond to her concerns about rusty orange tap water, thinning hair, and skin irritations in her home.

Edwards responded with citizen science. For preliminary data, he taught Walters to take tap water samples that he could test. Even though Edwards found exceedingly high levels of lead in Walters's water, he was initially ignored when he brought the findings to the U.S. Environmental Protection Agency. So Edwards created the Flint Water Study with his students, some funds from the U.S. National Science Foundation, and more funds from an online crowdfunding campaign. Participants in the Flint Water Study received special water vials, collected tap water according to a specific protocol, and mailed the samples for processing at Virginia Tech. Data were publicly available and displayed. Walters created the "Water Warriors" to collect samples and helped them use the data to support their political action.

The results of the project garnered national media attention and broader public pressure, forcing government actions (short-term provisioning of bottled water, testing of blood levels, movement towards long-term solutions) and inspiring community service (e.g., hundreds of union plumbers installed water filters for free). The project that began with data collection became a key element of a national political debate about social power in settings where technical expertise is necessary.

The third story of citizen science illustrates the achievement of "democratic" citizen science through the "contributory" style of citizen science. One way of understanding the relationship between the meanings of "citizen science" explored in this chapter is that the "democratic" definition represents a larger context in which the "contributory" style of citizen science resides. The lowest common denominator to citizen science projects is the collection and/or processing of data. From that focal point, the collaboration between scientists and nonscientists can expand. If the collaboration expands enough, the resulting new relationship then takes on the vision presented by Irwin, characterized by new perspectives, collaborative action, trust, etc., leading ultimately to societal influence shaping scientific agendas and norms.

Further Reading

Ballard, H. L. & Belsky, J. M. (2010). "Participatory action research and environmental learning: implications for resilient forests and communities." *Environmental Education Research* 16: 611-627.

Bonney, R. (1996). "Citizen Science: A Lab Tradition." *Living Bird* 15(4): 7-15.

Bonney, R., Ballard, H., Jordan, R., McCallie, E., Phillips, T. Shirk, J., & Wilderman, C. C. (2009). *Participation in Scientific Research: Defining the Field and Assessing Its Potential for Informal Science Education CAISE Inquiry Group Reports*. Washington, DC: Center for Advancement of Informal Science Education.

Conde, M. (2014). "Activism mobilizing science." *Ecological Economics* 105: 67-77.

Cooper, C. B., Dickinson, J., Phillips, T., & Bonney, R. (2007). "Citizen Science as a Tool for Conservation in Residential Ecosystems." *Ecology and Society* 12(2): 11. http://www.ecologyandsociety.org/ vol12/iss2/art11

Cornwall, M. L. & Campbell, L. M. (2012). "Co-producing conservation and knowledge: citizen-based sea turtle monitoring in North Carolina, USA." *Social Studies of Science* 42: 101-120.

Epstein, S. (1995). "The construction of lay expertise: AIDS activism and the forging of credibility in the reform of clinical trials." *Science, Technology, & Human Values* 20: 408-437.

Irwin, A. (1995). *Citizen Science: A Study of People, Expertise, and Sustainable Development.* New York, NY: Routledge.

Littman, M., & Suomela, T. (2014). "Crowdsourcing, the great meteor storm of 1833, and the founding of meteor science." *Endeavour* 38(2): 130-138.

3

TEACHING STUDENTS HOW TO DISCOVER THE UNKNOWN

Robert R. Dunn and Holly L. Menninger

In the early 1400s, at the end of the Middle Ages in what is now Italy, when knowledge was being reborn, anatomists would read from an ancient Greek text while their assistants dissected a human body and pointed out its parts. If the body looked different from what was written in the thousand-year-old text it was seen to be mutant, deviant, wrong. No matter that the ancient Greek knowledge was flawed and many of the rather ordinary observations that were being made would have dramatically improved on what was known. It would take a major scientific revolution for anatomists to begin to actually observe and learn from dissections. The idea that more knowledge could be gained was a breakthrough. Looking back, it's shocking how hard it was for early scientists to figure out obvious things.

Not so long ago, we were reminded of those Italian anatomists when walking past a classroom in which undergraduates were dissecting cats. Around the world — but particularly in the United States, where as many as 79% of middle and high school biology teachers report using dissections in their classes—millions of cats, dogs, pigs, and other mammals, including thousands and thousands of humans, are dissected in anatomy classes. They are dissected in order to teach students—including all of those who will eventually operate on your body—how an average mammal, amphibian, or other body works.

One can discuss the merits or the morality of having students perform dissections. We won't do either, though we can probably all agree that we owe it to the people and animals whose bodies are being pulled apart to use them as effectively as possible. My (Rob's) grandfather donated his body to science, for example, because he couldn't get into medical school when he was younger and he thought that in death he might do what he had proven unable to do in life—put himself to good use in a medical school. But the medical school in which he ended up probably did not put his body to good use. How do we know this? Because the vast majority of medical school cadavers are used to teach students what is already known rather than to make new discoveries. The vast majority of cadavers are used for Middle Age science.

In many college anatomy classes, dead animals are handed out to students. A teaching assistant, perhaps overworked, underpaid, or poorly equipped for the role, discusses how the dissection should be done. Students perform various forms of butchery. Students label, identify, or remove the parts of the body on which the teaching assistant has told them to focus. The students are told, at least generally, how said body part works. More body parts are dissected. More knowledge is provided. The bodies are then thrown away in special trashcans. The students and instructors alike go home, thinking of job applications, other classes, a love interest, or happy hour. The whole process repeats itself with a new group the next morning. This is funny because it is the way nearly everyone who has taken a biology class learned anatomy.

If you took a more advanced anatomy class it was likely distinguished only by the larger diversity of things into which you cut.

We don't mean to minimize the hard work of students or teaching assistants. What we are criticizing, however, is that we seem to largely teach anatomy in exactly the same way that it was being taught at the end of the Middle Ages. Specifically, students look at bodies of animals, but are not encouraged in any way to make real observations. Instead, they are instructed to look for what is already known and then if it does not look quite right, depict it the way it "should" look. Where differences among bodies are noted, they are seldom measured. Even when measurements are taken, they are seldom recorded.

Now, you might think we are confusing things. At the end of the Middle Ages we were ignorant about the body. Simple measurements could produce new knowledge. Now we understand the body. Of course, you are right to note that difference — or you would be, except that we still don't understand the bodies of animals very well. The function of the appendix is under new scrutiny, for example, as is the stomach. In fact, when it comes to basic morphology, the sorts of things that can be measured by students in large classes, we haven't made much progress in the last hundred years.

How and why do intestines vary among individuals? How frequent are different deformations of particular organs? Are there tradeoffs between investment in one organ and in another? How common are rare or poorly understood mutations in the bodies of cats, pigs, or even humans? Such mutations are hard to study because of their rarity, but we dissect so many pigs, cats, and other animals that even something that occurs in just one in a million animals turns up somewhere in some class each year. What else could be studied? There are potential discoveries right beneath students as they look up at their teaching assistants or teachers, but we are training students to ignore them, to see the general story at the expense of the truth.

It is not only in dissections and anatomy where we still teach in this antiquated way. Walk into an elementary school, a middle school, or a high school and you'll note that lessons based on demonstration science are the norm in K-12 education, not the exception. Consider, for instance, the carnation and food coloring experiment, the bean in the jar, or combining chemicals A and B to produce puffs of smoke. In fact, demonstration science is so much the norm that we don't even call it demonstration science, we just call it science education.

In our view, this typical approach to science education is not only out of date — it's broken. If this seems harsh, it's because we mistakenly assume that the world is mostly known and understood. In truth, much of the world is not nearly as well understood as we'd like to believe. To take a simple example, most species on Earth have yet to be named; until species are discovered, documented, and named, we can't really know anything else about them. By extension, therefore, we know precious little about most of life. If we are going to train generations of students to help us understand this unknown life, this unknown world, we must do better than Middle Ages science education. Thanks to teachers, we are starting to.

For the last four years, we have developed Your Wild Life (yourwildlife.org) as a program that engages the public in the study of the biodiversity in their daily lives. We work in homes, backyards, neighborhoods, museums, and science centers. We ask the public to do wacky things like twirl a Q-tip in their belly buttons to collect microbes inside their navels; set out cookie crumbs as bait to collect ants; and take pictures of spirited, leggy crickets hopping around their basements. These citizen scientists make important observations and contribute data to authentic scientific research. And then they go further. We encourage them (through social media and our blog) to participate in the whole process of science—analyzing the data, making novel insights, developing new hypotheses, and collectively determining where we as a public science research team should take the research next. In this sense, our projects push the boundaries of typical models assigned to citizen science projects.

Initially, our work—led by a team of scientists and science communicators—was situated squarely within the domain of informal science education. We sought the participation of the public (be they eight or 80 years old, single volunteers or entire families) in their free time, outside of the science classroom. For our earliest projects—Belly Button Biodiversity and School of Ants, which focused on documenting the diversity of microbes in our belly buttons and ants in our backyards, respectively—we partnered with the North Carolina Museum of Natural Sciences, our state natural history museum, to recruit participants and collect data at Museum-sponsored outreach events.

Over time, as our online presence and the number of Your Wild Life projects grew, our audience and participant base expanded widely. For example, for Wild Life of Our Homes, an investigation of microbial biodiversity on the surfaces in our homes, we recruited over 1,400 participating households, representing every American state and the District of Columbia. We are now a sort of digital museum of public science ourselves.

It was no surprise, therefore, that some inspired and persistent teachers, those practitioners of formal science education, took notice. They reached out to us and inquired about lesson plans to accompany our citizen science projects. These teachers wanted to abandon the carnations and the baking soda volcanoes in order to bring real scientific research to their classrooms. At the time, we had neither the content nor the expertise for creating such things.

So we embarked on developing a couple lesson plans based on Belly Button Biodiversity and School of Ants, but there was a problem. We didn't really know how to make lesson plans that would support the required curriculum standards and the teachers. In particular, many teachers weren't fully comfortable with science, particularly science where the answers remained unknown. In other words, the teachers couldn't pull together citizen science lesson plans on their own, and neither could we. We needed each other and we needed to work together, iteratively, to produce lessons in which the goal was not to reveal what we already know, but instead, create new knowledge. Because scientists have been largely avoiding K-12 education for the last five hundred years and teachers have mostly been avoiding science, the rift between our perspectives and fields was, at times, large. What seemed easy became both hard and interesting.

We persevered and our early efforts are now available for download at studentsdiscover.org. A lingering issue from this exercise was how to scale up our successes from the first few classrooms. This is the stage at which we caught a lucky break. A coalition of education researchers and experts in teacher professional development brought us in as partners on a large, ambitious project. *Wouldn't it be great*, we collectively thought, *to use citizen science as a means for engaging students across North Carolina and beyond in authentic research, to create opportunities for middle school students to make real scientific discoveries?*

Our project, "Students Discover: Improving Middle School STEM Outcomes through Scaling Citizen Science Projects," was funded by the National Science Foundation in 2013 and hit the ground running with its first cohort of middle school teachers in July, 2014. Rather than apply a lesson plan framework to pre-existing citizen science projects (as so often is the norm), Students Discover tries to flip the model. We partner early-career scientists with middle school teachers to co-create citizen science projects and associated lesson plans aligned to both state and national standards. We aim to bring real research—by which we mean novel and authentic question-driven research—to each participating classroom. We envision teachers and students transforming from observers and recorders to research collaborators and co-investigators, studying the ecology and evolution of species in our daily lives.

In its first year, the Students Discover scientist-teacher teams co-created projects that studied the tiny mites that live on our faces, the beneficial soil microbes that help a common weed thrive, fossilized shark teeth in ancient ocean deposits, and mammal diversity in urban habitats. In the second year, projects were added on backyard ants and their pathogens. In addition, we began to develop lesson plans for one hundred additional citizen science projects, with the goal of providing teachers access to as many projects as possible. These projects are now being implemented in classrooms across North Carolina, and the first pieces of student-generated data and discovery are rolling in.

The act of co-creation is time-consuming, labor-intensive, and messy, much like the process of science itself. Yet we think the payoffs, if we can achieve them—deep changes in teacher knowledge and instructional practice, increased student engagement in science learning, and improved science achievement—will be totally worth it. We will not stop working until real science—investigations where the answers are neither known nor predetermined—becomes the norm in the middle school science classroom.

And as for the dissections? We aren't quite there yet, but we have an idea of what could be done. We would have students take real measurements along with high-resolution digital images of the animals, including humans, that they dissect. They would also take a tissue sample of each animal (this might need to occur before the animals were preserved, which would be harder, but possible). The images and measurements would be sent to a database where they could be compared with other samples of the same species, and the tissue would be shipped to a tissue bank. With the database, anyone could compare the features of animals to understand how much they vary. With the tissue bank, the genes associated with unusual features could be identified.

Citizen science has the potential to transform the classroom experience, shifting students from passively receiving knowledge to being active partners in a process of learning and discovery. With every moment in class, the students conducting dissections would be reminded that the body they are looking at is, like their own body, still imperfectly understood. This revelation—of the incompleteness of knowledge, and the role of students and teachers alike in advancing knowledge—took until the late Renaissance to discover. It is a discovery we would do well to build on as we consider how our own science and education might, together, be reborn.

Further Reading

Barberán, A., Dunn, R. R., Reich, B. J., Pacifici, K., Laber, E. B., Menninger, H. L., Morton, J. M., Henley, J. B., Leff, J. W., Miller, S. L., & Fierer, N. (2015). "The ecology of microscopic life in household dust." *Proceedings of Royal Society B* 282(1814): DOI 10.1098/rspb.2015.1139.

Beasley, D. E., Koltz, A. M., Lambert, J. E., Fierer, N., & Dunn, R. R. (2015). "The evolution of stomach acidity and its relevance to the human microbiome." *PLoS ONE* 10(7): e0134116. DOI 10.1371/journal.pone.0134116

Bollinger, R. R., Barbas, A. S., Bush, E. L., Lin, S. S., & Parker, W. (2007). "Biofilms in the large bowel suggest an apparent function of the human vermiform appendix." *Journal of Theoretical Biology* 249(4): 826-831.

Bonney, R., Ballard, H., Jordan, R., McCallie, E., Phillips, T., Shirk, J., & Wilderman, C. C. (2009). *Public Participation in Scientific Research: Defining the Field and Assessing Its Potential for Informal Science Education. A CAISE Inquiry Group Report*. Washington, DC: Center for Advancement of Informal Science Education (CAISE).

Dunn, R. (2015). *The Man Who Touched His Own Heart: True Tales of Science, Surgery, and Mystery*. New York, NY: Little, Brown.

Elizondo-Omaña, R. E., Guzmán-López, S., & De Los Angeles García-Rodríguez, M. (2005). "Dissection as a teaching tool: Past, present, and future." *The Anatomical Record* 285B: 11-15. DOI 10.1002/ar.b.20070

King, L. A., Ross, C. L., Stephens, M. L., & Rowan, A. N. (2004). "Biology teachers' attitudes to dissection and alternatives." *Alternatives to Laboratory Animals* 32(1): 475-484.

Oakley, J. (2012). "Science teachers and the dissection debate: Perspectives on animal dissection and alternatives." *International Journal of Environmental & Science Education* 7(2): 253-267.

PART TWO

4

WHEN CITIZEN SCIENCE MAKES THE NEWS

Lily Bui

Broadcasting, believe it or not, comes from farming.

In modern vernacular, "to broadcast" means to transmit information by TV or radio, but the verb's original definition meant "to scatter [seeds] by hand or machine rather than placing in drills or rows." It may come as a surprise to you that broadcasting has just as much to do with farming and media as it has to do with citizen science.

In 1792, Robert B. Thomas started the *Old Farmers' Almanac,* a periodical circulated widely and regularly to farmers. Still in publication today, the *Almanac* serves two important purposes. First, it acts as an objective reference for weather and astronomical predictions, sourcing its observations from the farming community. Second, it facilitates a space where the community can share advice, anecdotes, recipes, and more with each other. But what does this have to do with citizen science?

If you've made it this far into the book, you have probably formed an impression of what citizen science is. For the purposes of this chapter, we'll think of citizen science as public involvement in inquiry, discovery, and construction of scientific knowledge, typically in the form of data collection, classification, or documentation.

Conceptually, the *Almanac* is not far from how some citizen science efforts are built: it incorporates public knowledge into a larger corpus of information predicated on a scientific question. Also, the *Almanac* illustrates another important concept—that citizen science does not happen in a vacuum. Citizen science inherently cultivates community, and we can conjecture from the *Almanac* that communities often need a means of communicating with themselves and with each other. And that, dear reader, is where media comes in.

The Role of Media in Citizen Science

When I say "media," I'm referring to processes that facilitate the documentation, reporting, or construction of information. In this chapter, I'll use examples of "mass media" (TV, radio, major news outlets) and what I'll call "micromedia" (blogs, social media, etc.) in both digital and non-digital contexts. You'll see quickly, however, that the lines between both distinctions blur very quickly and very easily.

But before we dive into the purpose of media in citizen science, it's important first to discuss the purpose of media in science proper. When I first set out to do research for writing this chapter, my literature scan for publications about "purpose" and "media" and "science" rendered results about journalism and public relations. This seems like the perfect locus at which to begin our conversation.

In *Journalism, Science, and Society*, editors Martin W. Bauer and Massimiano Buchi bring together voices that provide a critical view of science and media. As one of those voices, Tim Radford, the former science editor of the *Guardian*, identifies "a crucial tension in the focus of the mass media — particularly papers — on seeking a good narrative rather than seeking to advance public education as scientists sometimes seem to expect." In a separate chapter, Claudio Pantarotto and Armanda Jori describe their conversation with a biomedical company in which the interviewee describes the company's view on media: "Communication has two strategic purposes: firstly, to attract donations [...] secondly, to maintain the image and reputation of the institute."

These views beg excruciatingly essential questions: Is science journalism, in effect, PR for science and scientists, or is there a distinction between the two? If so, where is the distinction? What — if anything — is the responsibility of media in terms of public education and understanding of science? How can media add value to the way the public discovers, accesses, and consumes scientific information?

The answers to these questions depend largely on how you look at the role of media itself, which requires peeling back an additional layer. Media theorist James W. Carey presents two different models of communication that might be useful in helping us think through this. One is a "transmission model," in which one party imparts information to another, like in traditional broadcast TV and radio. (This is also a typically unidirectional relationship between media and people.) The other is a "ritualistic model," which is defined by things like sharing, participation, association, and fellowship. Based on what we know about citizen science and its collaborative involvement of the public, Carey's ritualistic view of communication seems to lay along the line of best fit.

While traditional media relies on transmission (e.g., a large newspaper prints a cover story and distributes it to the masses), I propose that citizen science media could and should shift toward the ritualistic model (e.g., many disparate sources collaboratively construct the universe of citizen science media) to raise public awareness about science; induct participants in citizen science projects; and close the feedback loop between the public, scientists, and media outlets.

The Current Citizen Science Mediascape

In very broad terms, there are several forms of "media about citizen science" that have crystallized in recent years:

- Media reporting on citizen science as a field/practice;

- Citizen scientists producing media about their participation in projects; and

- Scientists and researchers writing about citizen science (both within and outside of academic publishing).

The largest volume of citizen science media content likely resides in the latter two categories. Let's begin, however, with the first, which is essentially mass media's *modus operandi*.

Mass Media

At the time of writing this, you can count on one hand the number of mainstream media outlets that regularly produce content about citizen science: *National Geographic*'s Citizen Science Education initiatives, *Discover Magazine*'s Citizen Science Salon, and *Scientific American*'s Citizen Science blog.

These publications play important roles in raising awareness about citizen science. Articles and blog posts capture opportunities to participate in projects, research findings and results, and interviews with project managers and scientists. Some publications are restricted to one channel of delivery (either print or online), whereas others, like *Discover Magazine*'s Citizen Science Salon (a SciStarter project), are cross-platform, spanning both print and digital.

The important thing to note, though, is that in most cases, mass media outlets are reporting on citizen science and transmitting information one way. Although it is possible for audiences to give feedback in the form of online comments or writing to the editors, the nodes in this model rarely extend beyond the reader and the media outlet itself. Then participating in the citizen science projects that these media report on usually requires navigating away to that particular project page, a process that is disconnected from the medium through which the original story was received.

Figure 1: Traditional Model of Science Media

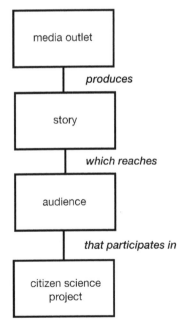

As yet, there is no such thing as a regular citizen science segment on television (in news or any other genre). Radio may be ahead of the curve, though, which I'll discuss in a separate section ahead.

Micromedia

Meanwhile, there has been an emergence of citizen science micromedia produced by non-professionals for smaller audiences. This mostly occurs online on blogging and social media platforms, and stems from a trend in the broader science journalism community, which strives to fuse a more direct and effective connection between scientists and the public.

In the blogosphere, sites like SciStarter and the Public Library of Science's (PLOS) CitizenSci blog produce regular stories about citizen science research, often authored by scientists, citizen scientists, educators, policy researchers, and others. SciStarter links readers directly to projects, which are organized in its database of hundreds of citizen science projects. The site also connects projects and project managers directly to media through various media partnerships. PLOS CitizenSci connects readers to citizen science projects and science papers. On PLOS CitizenSci, Caren Cooper, the Assistant Director of the Biodiversity Research Lab at the North Carolina Museum of Natural Sciences and former Cornell Ornithology Lab researcher, writes the "CitizenSci Scoop," a weekly blog series that highlights a host of topics in citizen science history and research.

Some citizen science platforms provide a forum in which participants convene and discuss projects. The exchanges in these forums can also be seen as a production of media or documentation. For example, the Zooniverse citizen science platform hosts a plethora of citizen science projects for fields spanning astronomy to climatology to archaeology. Each project has its own Talk site, where participants can talk to both each other and to scientists leading the projects, in case anyone needs clarification on how a project is running.

Public Lab, an open-source science community that focuses on the development of low-cost and low-tech sensor tools, encourages members of its community to contribute research notes to the Wiki-based site to document development, testing, and ideation efforts. The research notes are a cross between a blog post and scientific paper, which other members of the Public Lab community can comment on, share, etc. The notes can also be posted by anyone and everyone in the community.

Media also get produced for citizen science projects. #SnowTweets, a citizen science project that focuses on cryosphere research, asks citizen scientists to tweet observations of snowfall (even if there is zero snowfall). Their team then scrapes Twitter for these observations and uses the data for analysis. Project NOAH and iNaturalist are nature observation citizen science projects that invite participants to submit photographs of species sightings to their website.

Although no single source of micromedia reaches as many people as mass media can reach at a given time, the aggregate of all citizen science micromedia provides a more robust and representative view of what is actually happening within citizen science. Stakeholders should foster a multi-pronged approach to constructing the citizen science media discourse, and one camp that is accomplishing this very well is public media.

"Public media," as opposed to commercial media, refers to any media disseminated by or supported by a public broadcasting station or entity (e.g., PBS, NPR, BBC, etc.). Before I proceed, let's get one thing straight: I am an unabashed public media nerd. You also know by now that I am an advocate for citizen science. Despite my embedded biases, however, the convergence of these two worlds speaks to a much larger narrative—that of citizen science as a form of public participation in science and public media as a forum for public discourse.

Figure 2: Proposed Model for Citizen Science Media

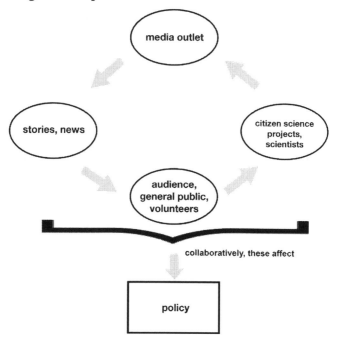

Public Media, Public Science

Here are some basic reasons why collaborations between citizen science and public media might be successful and even flourish:

Shared mission. Public media is a mission-driven industry with its roots in public education and pedagogy, which aligns with the many citizen science projects that offer educational value by design. Both public media and citizen science also seek to raise awareness about relevant issues and to engage people beyond consuming information alone.

National and local interests. Citizen science projects can illuminate very local issues (e.g., a dwindling frog population in southern Illinois) as well as national issues (e.g., climate change), while providing a means to engage with these problems. Public media often covers national and local issues, which can be aligned with relevant citizen science projects that advance a related field of research.

Infrastructure. Public media already has the infrastructure for disseminating content on air and online. Its organizing principle is also to create symbolic and material communities around public media content. Most citizen science projects constantly seek more participants and opportunities to build communities around citizen science.

Volunteerism. The heart of citizen science is the spirit of volunteerism, and many public media stations operate with the same credo. A citizen science project cannot be successful without volunteers to help collect data, and many public media stations rely on their volunteer networks to run events, promote programming, and in some cases even create content.

Believe it or not, public media and citizen science have already met and hung out, with many case studies underway (and an alphabet soup of station call signs to boot).

WHYY's The Pulse

Every few weeks, through a partnership between WHYY and SciStarter, producer Kimberly Haas features a citizen science project with some kind of connection in the Philadelphia area, the radio station's broadcast region. She interviews project managers, volunteers, and researchers about their work and encourages listeners to participate in the projects as long as the projects are active. These stories match with seasonal considerations. For example, a previous winter segment featured a citizen project called Tiny Terrors, which focuses on identifying the invasive woody adelgid. The story was framed by the context of winter in the Northeast, where the types of trees susceptible to the woody adelgid grow. *The Pulse* also runs stories on many other types of citizen science stories.

KVNF's iSeeChange

ISeeChange is a citizen science project and public radio experiment that ties citizen observations of weather directly into media. Participants submit weather observations in the form of photos, video, or tweets, which can be incorporated into news stories about climate change. This model opens up the newsroom and brings both public radio listeners and citizen scientists into the data-gathering and storytelling process.

WNYC's Cicada Tracker & Clock Your Sleep

The WNYC data news team ran two citizen science experiments involving sensor-generated data. The first, Cicada Tracker, involved having public radio listeners build an Arduino-based soil temperature sensor and submitting regular readings to the WNYC team via a simple app. The goal was to see whether there was a correlation between increases in soil temperature and the frequency of cicada sightings (there was). As people submitted their sensor readings, the data news team mapped the data points on a CartoDB map and also blogged about the process along the way. Similarly, the Clock Your Sleep project involved people with JawBone wearable sensors tracking their sleep patterns and self-reporting their sleeping patterns through a smartphone app. The data were collected, analyzed, visualized, and reported on via the WNYC website.

WCVE's Science Matters

WCVE Idea Stations has used their platform to connect their audiences to citizen science. They have featured regular blog posts, audio, video, and even live events that bring people to citizen science opportunities like FrogWatchUSA, the Great Backyard Bird Count, and Cicada Watch.

Science Friday

The educational arm of *Science Friday* seeks ways to repurpose existing radio content for education and the general public. *SciFri* has featured citizen science projects on its education page in the past and enlisted its audience to participate. Their latest project engages online listeners to tweet observations about the world using the #ObserveEverything hashtag, appealing to a historical tradition of observation-based science, which citizen science draws upon often.

PBS's NOVA Labs

NOVA Science is known for producing science documentaries for public broadcasting. Recently, they launched an initiative called NOVA Labs, which are citizen science projects hosted on the NOVA website. The various labs that citizen scientists can participate in span fields like solar energy, molecular engineering, cybersecurity, and more. In tandem with the companion educators' material (videos, interactive documentaries, and other content) already available through the site, NOVA Labs is meant to be a resource for educators, students, researchers, and citizen scientists alike.

In these examples, we see the potential to create more of a dialogue between news outlets and the public when it comes to scientific inquiry and investigation. Needless to say, I hope that public media does more with citizen science, and I hope that citizen science does more with public media, as there is great potential for mutual benefit in these collaborations.

The Big Picture

Because citizen science is still in its infancy, so too is citizen science media. However, a survey of the current mediascape shows that there are emergent trends and opportunities for experimentation as both the field and the subsequent coverage of the field grow. As media makers and communicators create the architecture for this space, it might be useful to think about how to design citizen science media with three important dimensions in mind: awareness, engagement, and impact.

Awareness: discovering, learning. How can citizen science media help raise awareness about local and national scientific discourse? How can citizen scientists themselves, as well as other stakeholders, be part of this process?

Engagement: understanding, knowing, sharing. What are ways that citizen science media can link people to opportunities to engage with news and citizen science projects? Who else can the media engage beyond the current pool of participants, and around what causes or issues?

Impact: doing, changing. What are the things that citizen science hopes to change in terms of public understanding of science, educational approaches, policy, etc., and how can media facilitate these efforts? How do we measure these changes, even design for them at the front end of media campaigns and citizen science efforts?

Media is, famously, the first rough draft of history." And because the emergent "ritual" model for citizen science media allows anyone to produce media about citizen science, the potential authorship of citizen science history is distributed more broadly than ever. As it stands, citizen science media and culture truly are what we actively and collaboratively make it.

Further Reading

Barrett, D. & Leddy, S. (2008). "Assessing Creative Media's Social Impact." White paper, The Fledgling Fund.

Bauer, M. W. & Buchi, M., eds. (2007). *Journalism, Science and Society: Science Communication between News and Public Relations*. New York, NY: Routledge.

Brumfiel, G. (2009). "Science journalism: Supplanting the old media?" *Nature* 458: 274-277.

Carey, J. W. (1989). "A Cultural Approach to Communication." In *Communication as Culture: Essays on Media and Society*. Boston, MA: Unwin Hyman.

Lewis, C. & Niles, H. (2013). "Measuring Impact: The Art, Science, and Mystery of Nonprofit News." Washington, DC: American University School of Education.

Lowe, G. F. & Martin, F., eds. (2013). *The Value of Public Service Media. RIPE@2013*. Goteberg, Sweden: Nordicom.

Walsh, L. (2010). "The Common Topoi of STEM Discourse: An Apologia and Methodological Proposal, With Pilot Survey." SAGE Publications.

5

SOCIAL MOVEMENT-BASED CITIZEN SCIENCE

Gwen Ottinger

On October 30, 2014, a coalition of environmental and community groups released a report entitled *Warning Signs: Toxic Air Pollution at Oil and Gas Development Sites.* *Warning Signs* reported on air sampling conducted by residents of communities in six states affected by hydraulic fracturing (fracking) and natural gas production activities. It integrated technical details of monitoring protocols and sampling results with first-hand accounts of the impacts of fracking on communities, resulting in an argument for expanded air monitoring by environmental regulatory agencies, full disclosure by natural gas companies of the chemicals used in natural gas production, and a "precautionary approach" to regulating the industry. The report was released on the same day that a peer-reviewed article about the study appeared in the journal *Environmental Health.* While the journal article did not include community voices directly, it defended the value of community-based monitoring and recommended including residents' participation in expanded government monitoring programs.

The well publicized, multi-community study is an example, on a grand scale, of one of two major variants of citizen science, described by Caren Cooper and Bruce Lewenstein as "democratic," in that it "address[es] the needs and concerns of citizens" and is "developed and enacted by citizens themselves" (p. 54). In contrast to Cooper and Lewenstein, this chapter argues that efforts like *Warning Signs* and AIDS treatment activists' work to change protocols for clinical trials are better characterized by their connections with larger political projects undertaken by what are known as "social movements." Social movements mobilize large numbers of people and organizations to raise an issue to prominence, change the way we think about it, and affect policy on the issue. In social movement-based citizen science, activist groups design studies not only to improve knowledge but to foster collective action and political change. Growing out of the environmental justice movement, the *Warning Signs* study brought together far-flung communities affected by unconventional oil and gas production to document the extent of the effects, highlight the government agencies' inattention to the problems, and further calls for stricter regulatory oversight.

Although few social movement-based citizen science projects are as extensive as the *Warning Signs* study, it nonetheless exemplifies several defining characteristics of this form of citizen science:

- Research questions grow directly out of the questions and concerns of citizen scientists;
- Credentialed scientists participate as allies, providing resources and advice without driving the research;
- Activists use innovative methods, including "do-it-yourself" (DIY) instruments;

- Questions and methods implicitly or explicitly challenge the adequacy of standard scientific approaches; and, as a result,
- Social movement groups engaged in citizen science face tradeoffs between scientific legitimacy and political efficacy.

Because social movement-based citizen science is by definition political, it is often discounted or dismissed by scientists concerned that it is not sufficiently objective to make a reliable contribution to scientific knowledge; policymakers, similarly, may believe that it is not rigorous enough to be responsibly used to inform policy. But its politics should not disqualify it as a contribution to science or policy. On the contrary, given that value judgments are inevitable in all scientific investigations, the explicitly political nature of citizen science grounded in social movements suggests ways that all forms of citizen science — and science in general — could become more robust by being more transparent and more deliberate about their own values. Furthermore, by diversifying the values that inform scientific inquiry, social movement-based citizen science can help scientists identify fruitful new methods and avenues of investigation.

Research Questions Are Community Questions

The first-hand accounts featured in *Warning Signs* make it clear that the health effects of natural gas production are of particular concern to nearby communities. In testimony after testimony, residents not only describe the symptoms they have experienced, they posit a connection between ill health and chemical smells. Caitlin Kennedy of Clark, Wyoming, for example, describes her experience: "It smelled like someone had turned a stove on without the pilot light on. I immediately got a headache, my nose started burning and I felt lightheaded." Similarly, in Pavillion, Wyoming, John Fenton reports "feeling tightness in my chest, nausea, throat irritation," and a number of other symptoms after being "overcome by a sickly sweet odor and an acid-like metallic taste."

Accordingly, the study is designed to probe the hypothesized connection between exposure to chemicals from natural gas production and illness. Its two research questions, "Are community members, workers, and animals being exposed to harmful airborne chemicals from fracking and other production activities?" and "Do the known health effects of those chemicals give cause for concern?" seek first to establish the presence in communities like Clark and Pavillion of chemicals that could cause illness, and then to connect them to potential health effects. The study's methods are also closely tied to the way that residents experience pollution: air samples were taken in places where community members had consistently noticed chemical smells or experienced health effects.

The defining role that citizen scientists' questions play in the *Warning Signs* study makes the project typical of social movement-based citizen science more generally. Their questions, moreover, are not motivated by curiosity alone; rather, they arise from local experiences that suggest a larger problem. In his study of a community-initiated health study in Woburn, Massachusetts—a common type of social movement-based citizen science known as "popular epidemiology"—sociologist Phil Brown describes the process of creating such projects as starting when individuals begin to notice both pollutants and health effects in their community; it continues with their hypothesizing a connection between the two and setting out to probe that connection.

Scientists Are Allies

Although citizen scientists define the research questions based on their experiences and observations, credentialed scientists still participate in social movement-based projects. The environmental groups leading the *Warning Signs* study recruited scientists to the project early on, asking them to review the monitoring protocol and help interpret data. The participation of scientists was also crucial in getting the study accepted to a peer-reviewed journal, although community leaders also routinely call on scientists even when they do not aspire to produce an academic publication. In Louisiana, communities that use bucket air samplers in their local campaigns, for example, ask local chemist Wilma Subra to put sample results in the context of reported releases from nearby petrochemical facilities.

Credentialed scientists play two important roles in social movement-based citizen science. First, they can improve study design, by helping citizen scientists make sure the methods that they've chosen are well suited to the questions they want to answer. In the *Warning Signs* study, the research team knew that they wanted to test air quality near compressor stations and had imagined using bucket air samplers to do so. Allied scientists, however, suggested that they would need a different instrument to measure the pollution they were interested in, so the study was modified to include monitoring with formaldehyde badges at those sites.

Scientists' involvement can also increase the legitimacy of science produced by activist groups. It can facilitate publication in peer-reviewed journals and coordination with regulatory standards, both widely accepted markers of scientific credibility. As spokespeople for social movement-based projects, scientists help increase the likelihood that decision makers will take citizen scientists' results seriously, by presenting the results from an authoritative, external perspective (a process that sociologist Abby Kinchy calls the "epistemic boomerang") and framing them in the technocratic language used in regulatory proceedings.

When choosing scientists to work with, community leaders are careful to find scientists supportive of their goals and methods. *Warning Signs* researchers, for example, looked for experts who not only could offer expertise on oil and gas production but also would acknowledge that community members themselves were well positioned to identify sampling sites and collect air samples. The "expert-activists" who participate face the challenge of protecting their scientific credibility while at the same time helping social movement groups achieve their political goals; in his research, sociologist Scott Frickel shows how they do so by forming under-the-radar networks of like-minded colleagues, which he calls "shadow mobilizations."

Methods May Be "Hacked"

The *Warning Signs* study reported on results from air monitoring conducted with "bucket" air samplers. The buckets are a kind of "grab sampler": they pull a volume of air into a container that is then sent off to a laboratory for analysis. Most grab sampling is done using a device called a Summa canister. Summa canisters are stainless steel spheres with a glass lining; they come from the lab already evacuated, and the entire device has to be sent back to the lab every time a sample is taken. Buckets, in contrast, house a bag made of special, nonreactive plastic called Tedlar in an ordinary paint bucket. Scientific-grade tubing is used to connect a miniature vacuum cleaner, available in consumer electronics stores, which evacuates the bucket and allows air to flow into the bucket. Once a sample has been taken, only the Tedlar bag has to be sent to the lab; the rest of the device remains in the community and can continue to be used.

Buckets are just one example of a scientific instrument used by citizen scientists to replace a standard scientific method or instrument with one that is more accessible to ordinary people. Since 2010, many other such instruments and techniques have been developed by members of an organization called Public Lab. Their website (publiclab.org) offers tools for measuring water quality, sensing chemicals in indoor air, and creating aerial maps of areas affected by pollution or natural disasters, to name few.

These DIY monitoring techniques share a number of characteristics. First, they are tailored to the questions of citizen scientists. The bucket's design, for example, is optimized for community concerns about very strong odors from petrochemical facilities that might last under an hour, but that might occur several days in a row. Buckets accordingly have a sampling period of several minutes — long enough not to miss the smell if the wind shifted for a moment, but short enough not to dilute the worst smells with cleaner air. The easily replaceable Tedlar bag allows them to always be at the ready in the community, instead of being unavailable while a sample is being analyzed, as Summa canisters are.

The level of precision offered by DIY monitors is also dictated by the questions and interests of users. Very small concentrations of air toxins can be harmful to human health; as a result, bucket samples are analyzed using a technique that can measure those chemicals at the parts per billion (ppb) level. In contrast, an indoor air monitor developed by Public Lab doesn't take numerical readings at all. Instead, it roams around a room, changing color when chemicals are detected, allowing users to locate the source(s) of hazardous chemicals in their homes.

DIY instruments tend to be built by users. In bucket trainings, community members build buckets together, in addition to learning how to use them. Public Lab posts parts lists and instructions on its website for the instruments developed by members. Because of the hands-on nature of these devices, innovations and improvements also come from users. Buckets no longer use actual buckets to house their sampling bag; instead, they use similarly sized clear plastic storage containers commonly used for bulk foods, a user-inspired modification that solves a number of problems posed by the original design. This kind of modification is actively encouraged by Public Lab's "open source" ethic, in which creators of a technology don't claim intellectual property rights but instead make their creation freely available to others to use, modify, and distribute. Because they are constructed with readily available parts and common household items, these instruments also tend to be much less expensive than their standard scientific counterparts: buckets cost under $100 to build, whereas Summa canisters cost $500 or more to buy.

Scientific Critiques Are Central

Alternative instruments and methods should not be seen as merely science on the cheap. Rather, they enable another fundamental aspect of social movement-based citizen science, that of critiquing the way that science is ordinarily done and changing standard practices. Participants in the *Warning Signs* study took air samples near fracking sites to call out scientists for their neglect of the air quality impact of fracking in places where people live and where children go to school. This criticism—that the effects of petrochemical pollution on neighboring communities are systematically neglected—is routinely leveled at regulatory agencies by bucket users around the world, with the effect that, nearly 20 years after buckets were first developed, federal regulation now requires fenceline monitoring at all U.S. oil refineries.

But citizen scientists don't only want to push scientists into doing research that is being left undone; they also advocate for changes in scientific methods and standards of evidence. In his work, the environmental health researcher Phil Brown explains how popular epidemiology challenges the ways that professional public health researchers decide whether there is enough evidence to conclude that an environmental hazard is affecting people's health. The public health standard is to prefer false negatives over false positives — that is, they choose to err on the side of saying there is no problem when there is, rather than take the risk of declaring that there is a problem when there actually isn't. They insist on a very high level of statistical significance; they want to be 95% sure that any elevated rates of disease that show up could not have occurred randomly. Popular epidemiologists, in contrast, argue that they should not have to meet these strict standards before the government will intervene to protect their health, and they push for a more precautionary approach with a lower bar for statistical significance.

Bucket users also challenge scientific standards for establishing what constitutes an environmental health problem. Environmental regulators take air samples over a 24-hour period and compare them to health-based standards set by the government for air quality, sometimes averaging a whole year's worth of samples to compare to an annual average. Bucket users, working with individual 5-minute samples, focus their analyses on chemicals frequently detected by buckets and compare them directly to whatever standards are available, whether they are for 8-hour, 24-hour, or annual averages. In doing so, they highlight the absence of any standards for some chemicals known to affect human health; they point out the enormous uncertainties in the standards, which can differ by as much as ten or even 100 times from one set of agency guidelines to another; and they assert that relatively short-term exposures, not just long-term average exposures, can directly impact people's health.

Scientific Legitimacy Traded for Political Efficacy?

Despite their critique of standard scientific methods, the citizen scientists involved in the *Warning Signs* study made several moves to establish their scientific legitimacy. They partnered with credentialed scientists and they published their findings in a peer-reviewed journal as well as in an activist report. Furthermore, the article and report argue for the legitimacy of sampling results by drawing on the authority of the U.S. Environmental Protection Agency (EPA): both refer to the EPA federal reference methods used to analyze samples, and the report describes how the sampling followed "stringent quality control protocols originally designed with EPA Region 9."

Establishing scientific legitimacy is important for citizen science to actually help bring about the changes that social activists seek. Bucket results are taken seriously by regulators, at least to a limited extent, because air samples are analyzed in certified labs, using a method established by the EPA. As a result, bucket samples showing very high levels of chemicals can lead regulators to look for the source of those elevated levels and compel facilities to clean them up. Similarly, Shannon Dosemagen and other founders of Public Lab report in an article that the participatory mapping practices it helped foster in disaster-affected communities were of great interest to international development agencies. These agencies use maps to guide and organize their own, top-down activities, and the agencies' interest in geospatial data lent legitimacy to community-generated maps.

Scientific legitimacy, however, may come at a cost: where social movement-based citizen scientists align themselves with expert practices for the sake of scientific legitimacy, their critiques of standard scientific practices are apt to get lost. International development agencies insisted that Public Lab condense their work with communities into very short time periods, attempting to impose a standard view of mapping as a straightforward technical procedure—undermining the organization's goal of fostering a deeply participatory practice that could represent locals' place-based knowledge. And bucket users' desire to see regulators act on air quality data that they collect has led them to push for more and more real-time air monitoring, which offers the possibility to generate long-term averages—and the potential for scientists to ignore the episodic peaks of pollution that buckets measure.

In conducting citizen science, then, social movement groups must constantly make tradeoffs between increasing the scientific authority of their claims and amplifying their critiques of science as it is usually practiced. In her book, Abby Kinchy shows the consequences of this tradeoff very clearly in discussing two different community-led approaches to showing that genetically modified maize was finding its way into supposedly unmodified maize through cross-pollination, contrary to seed company scientists' arguments that it would not happen. One of the groups focused on genetic changes in the seeds, in keeping with well-established science, which holds that it is in the genes that any unintentional cross-breeding will be visible. The other extended their investigations to look at mutations in the plants themselves, taking seriously locals' implicit hypothesis that genetic cross-contamination would be visible in the crop, but sacrificing their credibility with scientists.

Politics Don't Detract from Contributions to Science

Because social movement-based citizen science produces knowledge with an explicit goal of creating social change, critics charge it with "bias." Even those who might be sympathetic to citizen scientists' causes are tempted to regard their data gathering activities as political, not scientific, endeavors. This view draws on the widely held but erroneous belief that scientific investigations led by credentialed scientists are free of social values—except in anomalous cases involving corrupt scientists.

In reality, *no* science is devoid of social values. Even scientists who are scrupulously observing the standards in their field and taking every precaution to ensure that their personal desires or prejudices do not color their research must still make numerous decisions about how to do their work. What questions should they investigate? What terms and metaphors should they use to talk about their findings? Should they err on the side of seeing something that isn't there, or missing something that is? As philosopher Kevin Elliot shows, values necessarily enter into all of these decisions. Scientists should not be expected to eradicate values from their research, although they should of course be prohibited from manipulating data to suit their political purposes. Instead, both Elliot and fellow philosopher Heather Douglas believe that science would be strengthened if scientists were to be more reflective and transparent about their values. Doing so, they suggest, would permit a more productive societal conversation about the limitations of our knowledge, the implicit biases that we might want to correct, and the alternative ways of looking at things that might be more in keeping with our shared values.

Viewed in this light, the explicitly political nature of social movement-based projects can actually help make science more robust and more responsive to social needs.

Unlike mainstream scientists, activists are transparent about the values that inform their scientific activities. And in their critiques of standard scientific practice, they call attention to implicit value judgments being made by scientists that might not seem appropriate when viewed from other perspectives. The high level of statistical significance that epidemiologists insist on, for example, could well be deemed unreasonable by a society that thought it right to err on the side of caution when it came to protecting people's health.

Of course, just as there are cases where credentialed scientists manipulate data to prove a point in which they are personally or politically invested, there may be instances in which social movement-based citizen scientists manipulate data to serve their interests — and such behavior is just as unacceptable in social movement-based citizen science as it is in mainstream science. But it is also just as anomalous. The political agendas of social movement groups do not make them any more susceptible to corruption; if anything, citizen scientists who want to challenge accepted approaches are likely to be even more committed to making sure their methods are beyond reproach.

Responsibly conducted, the science conducted by activist groups can make a contribution to scientific knowledge writ large, not only by helping to amass data that credentialed scientists do not or could not collect. It can also diversify the values that inform scientific research and prompt discussions of what values ought to be informing scientific practice. By doing so, this kind of citizen science could help to identify new and fruitful areas of inquiry, and approaches to pursuing them that have the potential to be of greater benefit to society.

Summary

Social movement-based citizen science is characterized by a number of features that distinguish it from citizen science in which citizens participate in the research projects of academic scientists. It starts from citizen scientists' questions and hypotheses, with credentialed experts playing a supporting role. It invents alternatives to standard scientific instruments and methods and, through its novel approaches to community-based questions, critiques mainstream science with an eye toward change. Activist groups must decide how to trade off the political influence that comes with scientific legitimacy against their ability to have their critiques recognized and responded to in mainstream scientific practice. Finally, while running the risk of being dismissed as "unscientific" for its explicit political commitments, social movement-based projects actually have the potential to strengthen scientific knowledge by making the values inherent in *all* scientific research more transparent, enabling those values to be adjusted in keeping with societal needs and priorities.

Further Reading

Brown, P. (1992). "Popular Epidemiology and Toxic Waste Contamination: Lay and Professional Ways of Knowing." *Journal of Health and Social Behavior* 33: 267-281.

Coming Clean and Global Community Monitor (2014). *Warning Signs: Toxic Air Pollution Identified at Oil and Gas Development Sites.* http://comingcleaninc.org/warningsigns

Dosemagen, S., Warren, J., & Wylie, S. (2011). "Grassroots Mapping: Creating a Participatory Map-making Process Centered on Discourse." *Journal of Aesthetics and Protest* 8.

Douglas, H. (2009). *Science, Policy, and the Value-Free Ideal.* Pittsburgh, PA: University of Pittsburgh Press.

Elliot, K. (2011). *Is a Little Pollution Good for You?* Oxford, UK: Oxford University Press.

Kinchy, A. (2012). *Seeds, Science, and Struggle: The Global Politics of Transgenic Crops.* Cambridge, MA: MIT Press.

Macey, G. P., Breech, R., Chernaik, M., Cox, C., Larson, D., Thomas, D., & Carpenter, D. O. (2014). "Air Concentrations of Volatile Compounds near Oil and Gas Production: A Community-based Exploratory Study." *Environmental Health* 13: 82.

Ottinger, G. (2010). "Buckets of Resistance: Standards and the Effectiveness of Citizen Science." *Science, Technology, and Human Values* 35: 244-270.

6

CITIZEN MICROBIOLOGY: A CASE STUDY IN SPACE

David Coil

At 3:25 PM on April 18, 2014, I stood on the viewing platform at Cape Canaveral, Florida, watching a massive rocket carry a nationwide citizen science microbiology project into space. This project would catalog hundreds of types of bacteria living on the space station, survey thousands more bacteria from participants around the country, and measure the growth of common bacteria in space. Mixed with the excitement and relief was a feeling of amazement that we live in a time where such things are possible. New and cheaper technology has completely changed our understanding of microbiology in the last decade or two. We can relatively cheaply ask questions that weren't even conceivable in the recent past. These changes, along with rapidly growing public interest in microbiology, have created the perfect conditions for an explosion of what we call "citizen microbiology." Our project involving microbes in space is but one example of this new and exciting field.

Over the last decade, microbiology has seen a renewed surge of interest in popular media, books, and films.

While some of this relates to topics such as global pandemics and new diseases, increasing attention is being paid to subjects like the importance of beneficial human-associated microbes or the problem of antibiotic resistance. Given the current level of public interest in both microbiology and citizen science, it is perhaps no surprise to hear that citizen microbiology is taking off. In this chapter I'll discuss the idea of citizen microbiology, the opportunities and challenges therein, a few examples, and one detailed case study.

Before going into the details of citizen microbiology, a few definitions might be in order. A "microbe" is traditionally defined as a living organism too small to be seen with the naked eye. For our purposes this includes viruses, bacteria, fungi, and various other tiny creatures. "Microbiology" is the study of microbes and a "microbiome" is the collection of microbes found in a particular habitat (e.g., on a person or in a house).

A few microbiology citizen science projects that involved culture-based monitoring (i.e., growing microbes on plates in the lab) go back decades. For example, the State of the Oyster project in Washington State has helped volunteers monitor edible shellfish populations for harmful bacteria since 1987. However, the ease and low cost of DNA sequencing has been a major force for change. The majority of citizen microbiology projects today are less than ten years old, and rely in some way on cheap and easy DNA sequencing. This sequencing allows researchers to quickly and accurately identify most of the microbes in a given sample.

Microbes are hard to see, often viewed negatively, and have large impacts (good and bad) on human health. These features create both opportunities and challenges in conducting citizen microbiology.

Opportunities & Challenges of Citizen Microbiology

Every human being carries a complex and unique collection of microbes, making each person a valuable data point in understanding the human microbiome. Given our increasing understanding of the critical role played by microbes in human health, this understanding may transform numerous aspects of healthcare at an individual level. In addition to human-associated microbes, citizen microbiology efforts involving environmental and water monitoring can be extremely helpful in understanding microbial ecology.

Beyond the scientific benefits, there is a tremendous educational opportunity with microbiology. Many people react negatively to words like "microbes" and "bacteria." It is far more common to find the term "germs" in the media, usually portrayed in a health-influencing and negative way. Engaging the public through actually doing microbiology provides an opening to discuss the fact that microbes are everywhere, and the vast majority of them are harmless or beneficial. Increased awareness of this fact has important implications for human health, both directly (e.g., through reduced use of unnecessary antibiotics) and indirectly (e.g., shifting away from "kill all the microbes" that is probably counterproductive for health).

Another opportunity with citizen microbiology is the accessibility of samples: to get started all you often need is a sterile swab. Citizen microbiology can also be adapted in a hands-on manner in the classroom in a way that might be difficult with, say, endangered birds. An excellent example is the Phage Hunters project run by Graham Hatfull at the University of Pittsburgh, where students actually discover and characterize novel bacteriophages (viruses that infect bacteria).

Citizen microbiology also presents a number of challenges, some of which are shared with other citizen science projects but many of which are unique to, or more problematic, when dealing with microbes. There is often, for instance, a very strong negative association with microbiology and microbes. This is really both a challenge and an opportunity, since educating the public about microbiology should be a primary goal of any citizen microbiology project. Many people are both surprised and interested to learn how important microbes are to the world around us and our own health. In our experience, many times all that is needed is a couple examples of how "germs" aren't all bad to get people to be more open-minded about microbiology.

For the researchers, the logistics of organizing sample collections with citizen scientists can be quite complex. For example, samples collected for DNA analysis need to be protected from contamination and often kept frozen or otherwise preserved. This can be particularly difficult in the absence of electricity, which would require suboptimal chemical preservation methods or lugging around crates of dry ice. Human-associated microbes run into issues related to privacy, informed consent, and human-subject research. Solutions to this problem range from pretending it doesn't exist, to anonymizing all data, to (for example) collecting microbes from a cell phone instead of a person directly. Actually growing microbes as part of citizen microbiology (or in an educational setting) can present biosafety concerns. When microbes are given rich growth conditions (lots of food, warmth, liquid, etc.) it can be hard to predict what will appear. In particular, growth of human-associated microbes typically requires specialized equipment and training to ensure a minimal risk of either contamination or spread. Government regulations, and transportation/collection permits are other potential snags. In one frustrating example from our own lab, we recently discovered that while we could have mailed animal feces (rich with microbes) internationally without permits, once we had extracted DNA from the same samples it was considered highly regulated "biological material from a protected species."

Beyond the considerations in the field, one of the challenges with citizen microbiology — particularly that associated with humans — is in not over-interpreting the data. Conversations about the human microbiome tend to range between "kill all the germs" and "I take three kinds of probiotics and am considering a fecal transplant to get a more healthy microbiome." Scientists involved in citizen microbiology need to be very careful about how they present information about the human microbiome. Along these lines, there is a lot of concern about "self-experimentation" with projects that measure the microbiomes of participants. There's nothing to prevent people from radically changing their diet or lifestyle just to see what that does to their microbiome. The problem is mainly with interpretation: surely, for instance, if you eat nothing but beets for two weeks you'll observe changes in your gut microbiome, but no one can really say (yet) what those changes mean.

Another challenge is that of communicating the data back to the public. Traditional outputs of bacterial surveys include statistics and graphs (with dozens of Latin species names) that are hard to make sense of. Finding ways to display this complex data in a way that is meaningful to the public is, to my mind, one of the great remaining challenges in citizen microbiology.

Citizen Microbiology in Action

Most current citizen microbiology projects are focused on low-cost DNA sequencing to ask questions about what microbes are living where, and they collect data in collaboration with the public. For example, the Wildlife of Our Homes project examines what microbes (and other organisms) are present in the homes of volunteers. The Home

Microbiome Project went even further and found people who were about to changes houses, sampling both houses before and after—as well as the participants themselves—in order to understand the relationship of the human microbiome and the home microbiome. People are often very excited to participate in this kind of project, as they not only have the opportunity to learn more about microbiology in general, but also to learn what lives in their own home. Who isn't curious about that?

However, the closer a project gets to the participant (e.g. a local beach, versus your home, versus you) the more potential legal, ethical, privacy, and biosafety complications arise. Entering this realm are projects where members of the public can collect personal samples from themselves (skin, saliva, feces, etc.) and have the microbiomes of those samples analyzed. Two example projects in this area are the publicly funded American Gut Project and the privately funded uBiome Project. These projects, along with conventional microbiology research (e.g. the Human Microbiome Project) are already sparking a paradigm shift in our understanding of human health and disease and our interdependence with our microbes. This is a case where public participation can generate critical scientific data, as well as be directly relevant to the participant. As discussed above, this presents a number of opportunities as well as challenges.

I have experienced many of these opportunities and challenges though my involvement in helping to organize a nationwide citizen microbiology project called Project MERCCURI. MERCCURI is a tortured acronym for Microbial Ecology Research Combining Citizen and University Researchers on ISS (International Space Station). This project had several interrelated goals:

- To conduct a large nationwide survey of microbes found on shoes and cell phones.

- To collect microbial samples from the International Space Station.

- To observe the growth of a number of "non-pathogenic" (non disease-causing) microbes in microgravity on the ISS and compare this to growth on earth.

- To use these three scientific goals to engage the public in thinking about microbiology, and to a lesser extent, doing science in space.

This project was conceived by and co-organized with Science Cheerleader, a nationwide organization of professional cheerleaders pursuing science careers. Through this project, we organized a number of events at various venues, usually sporting events or museums. At these events, members of the public volunteered to swab their cell phones and shoes for microbes. These swabs were later analyzed to determine which bacteria were present there and for comparison to similar swabs on the ISS. Also at these events, different swabs were taken from surfaces like doors, handrails, etc. Bacteria from these surface swabs were cultured, and a candidate species was chosen from each event to fly to the International Space Station for the growth experiment.

Over the course of the project we experienced many of the challenges and opportunities common to large citizen science projects. A number of additional challenges and opportunities arose because of the microbial component. We present here a brief summary of our experiences in the hope that it is helpful to anyone participating in, organizing, or simply interested in citizen microbiology.

For Project MERCCURI, we dealt with the biosafety consideration by only having members of the public collect swabs, but not be involved in growing the organisms (all of which took place in a microbiology laboratory at the University of California Davis). Swabbing a doorknob probably presents less risk than touching it normally! We dealt with some regulatory issues by limiting the project to the United States. Most importantly, this meant that samples couldn't get stuck in customs for days, killing the microbes or confounding the results.

To address privacy concerns, all participants signed a detailed consent form. We also had a separate photography consent form, particularly if minors were involved. Privacy was addressed through barcoding the samples and keeping participant information separate from the samples themselves. We also agreed to pool the data from each event, and not track individual participants for this reason. Because of the pooling, and the fact that we didn't collect samples from people directly, we were able to get approval for a waiver from an Institutional Review Board (IRB). If, for example, we had given each participant data about their own microbes (as with uBiome and American Gut), this could have become much more complicated. IRB approval is normally required for any human subject research at any publicly funded institution.

Sample preservation was addressed through the use of dry swabs (freezing not required) and giving event coordinators a FedEx account number so that all samples could be shipped overnight to the lab at UC Davis. This neatly avoided the biosafety issue of growing microbes on site, but did require that participants were on the ball. In one unfortunate case a group of volunteers lost the sterile swabs we mailed and bought cotton swabs at a local drugstore, which turned out to be heavily contaminated with fungal spores. In several cases, swabs were left in hot car trunks for a couple of days and didn't produce any living microbes by the time they got to California.

Logistical and organizational constraints aside, our biggest challenges related to communication about the project. First, even explaining the project to people was challenging, given the many moving parts and the non-obvious relationship of cheerleaders, microbes, and space. Second was the major hurdle of the reaction of many participants along the lines of "germs are gross" or "I'll bet you'll find a lot of nasty stuff on my cell phone." Part of how we dealt with these two challenges was through providing access to relevant information, including a website with information about the project and information fliers distributed to everyone we talked to about the project. Once the candidate species were selected for flight into space we created "baseball cards" of each microbe that emphasized the beneficial (or at least not harmful) nature of all the bacteria we chose.

But anecdotally, our biggest success with regard to public education was simply through talking to hundreds and hundreds of people at these events. The very nature of the project drew people to our tables and attracted volunteers who might not otherwise have given microbiology a second thought. People were excited to participate in a nationwide survey of microbes and many were thrilled at the chance to be involved with something associated with space. Through these "hooks" we were able to convey our core messages about the ubiquity and benefits of microbes.

Conclusion

As discussed in previous chapters, citizen science is an incredibly powerful tool from both the perspectives of scientists and the public. Scientists gain the benefits of additional data and samples, as well as the opportunity to educate people about their work. Participants gain a chance to contribute to the process of science and to learn and become excited about a particular area of science. Citizen microbiology shares much with other kinds of citizen science projects, but brings some unique challenges and opportunities. Challenges include negative associations with microbes, logistical issues, privacy concerns, and problems with both interpretation of data and communication of the results. The opportunities include the ease of many experiments, the potential value of the data, and getting people excited about the microbes that affect the world around us and our own health. Because of the existing preconceptions about microbes (both good and bad), and the possible human health implications, citizen microbiology has incredible potential on both the scientific and educational sides of the coin.

Further Reading

Project MERCCURI: http://spacemicrobes.org

State of the Oyster: https://wsg.washington.edu/state-of-the-oyster-study-testing-shellfish-for-health-and-safety-2

Phage Hunters: http://phagesdb.org/phagehunters

Wildlife of Our Homes: http://homes.yourwildlife.org

Home Microbiome Project: http://homemicrobiome.com

American Gut: http://humanfoodproject.com/americangut

uBiome: http://ubiome.com

CONCLUSION

THE AGE OF CITIZEN SCIENCE

Eric B. Kennedy and Darlene Cavalier

The age of citizen science is upon us. As has been illustrated in this volume, this transformation means more than simply a new kind of volunteer labor. The age of citizen science heralds the potential of a fundamentally different relationship between scientists and the public, and between researchers and the questions they ask.

In its simplest form, citizen science challenges the norms of who ought to be welcomed into the world of science. Where experts have traditionally dominated the scientific enterprise, the citizen science movement casts forward an alternative vision: that the average citizen ought to be able to participate as well. The form of participation can vary dramatically, ranging from collecting data on an issue of passion through to fundamentally reorganizing the kinds of questions and projects at hand.

Taken a step further, however, citizen science advocates are arguing implicitly and explicitly for a radical change to the structures of political power. The call for citizens to be involved much more directly expands quickly to challenge norms in education, emerging technologies, and government decision making. Citizen scientists argue that classroom experiences—long driven by the "sage on the stage"—and formal degrees need not be factors in determining who can ask research questions and who can gather and interpret the data. Others advance the worlds of do-it-yourself, hacking, and open source hardware, challenging the conventional wisdom that only professionals can access or build the instruments needed to collect, analyze, and share data. More still push back on the notion that government representatives are able to authentically and completely represent their citizens, suggesting instead that the public needs to take a much more direct and active role in shaping decision making. Citizen science is about much more than just the scientific process itself—it's a fundamental reorienting of power relations among citizens, government, and society as a whole.

To understand citizen science fully, however, requires more than just understanding this big-picture context. It necessitates a careful examination of how we all found ourselves here; a better understanding of the participants' and project owners' motives; ongoing research into the intended and unintended outcomes of projects; and long-term, transdisciplinary collaborations to identify and embrace those communities which have not been represented thus far.

Our Story of Citizen Science

Some of this history is best understood through the personal narratives of citizen scientists. In Darlene Cavalier's chapter, we begin by closely examining one person's journey into the world of science engagement. She encounters and confronts traditional norms, including the exclusion of particular groups, stereotypes about what a scientist looks like, and assumptions about how much the public really cares about science. Her journey, however, highlights some intriguing possibilities: that there is an emergent desire to reform these antiquated notions, and to empower new people to engage with the scientific enterprise in a new way.

We then shift from examining a single tree to looking at the forest as a whole: the state of citizen science and how it relates to governments and policy. Eric Kennedy illustrates the ways in which citizen science must be understood in the context of its historical and political landscape, arguing against narratives that present citizen science as either a brand new movement or ubiquitous throughout history. Instead, it's a push that takes longstanding questions of legitimacy, participation, and engagement and reframes them around the public demand to engage directly in the inner workings of science itself.

The story underpinning the emergence of citizen science, however, has hardly been a simple one. As Caren Cooper and Bruce Lewenstein illustrate in their chapter on the history of citizen science, the movement interweaves several distinct strands of theory, practice, and purpose. Citizen science, they argue, can take on several meanings depending on its historical context, each of which comes with its own emphases and focuses. Whereas "participatory" citizen science emphasizes the importance of including the public in the collection and analysis of data, "democratic" citizen science offers a more fundamental questioning of what ends science ought to work to achieve. As such, it's important to understand that citizen science isn't a homogenous category, but rather can take on different forms depending on its purposes, methods, and normative commitments.

For Robert Dunn and Holly Menninger, both of these understandings of citizen science challenge more than just science: they mean that we must rethink education as well. They advance an argument against the practice of rote memorization in classrooms. Not only is it poor pedagogy, they argue, but it may well mean that we are missing an incredibly valuable source of information: the endless trove of research being conducted by students as they dissect bodies, conduct experiments, or even just wander in their communities. Through their experiences of bringing together scientists and teachers, Dunn and Menninger argue that citizen science can empower student to learn more effectively and contribute new knowledge to important questions.

The chapter by Lily Bui also exemplifies this kind of broader challenge by the citizen science movement: not only to science itself, but also to the education, media, and decision making that surrounds it. Through several examples, Bui illustrates the way citizen science is incompatible with the traditional one-way flow of established media. Citizen science requires a dialog between the public and scientists, wherein ideas, inspiration, and learning can be readily exchanged between the two groups.

Environmental examples offer some of the sharpest illustrations of these many challenges coming together. Through the case study of public engagement in energy and pollution production, Gwen Ottinger shows us that citizen science offers the public a meaningful way to affect decisions that matter to them. Ottinger doesn't shy away from the challenges and shortfalls of these kinds of projects, ever attuned to the socio-political dimensions that enable and constraint public actions. This offers a glimmer of what could be a way for citizens to engage when they've long been systemically excluded from matters constitutional to their very health.

Based on his experience with Project MERCCURI, David Coil offers a vision into the cutting edge of citizen science. While some kinds of science were much more accessible to public intervention over the past few decades, argues Coil, it's only with advances in technologies and opportunities that microbiology is beginning to see this kind of growth. These advances are unlikely to end anytime soon, he suggests, as the advent of wearable technologies, constant health monitoring, and access to big data offer endless new potential for citizen engagement. Indeed, it would seem as though microbiology—along with many other fields, like space exploration, big data, and nanotechnology—mean that the expansion of citizen science is only just beginning.

The Future of Citizen Science

Through the preceding chapters, this volume illustrates the incredible variety of citizen science projects. Such endeavors can be found in nearly every field, innumerable locations, and working toward almost any ends. (See, for instance, the *Journal of Microbiology & Biology Education*'s March 2016 special issue on citizen science.) Citizen science is prolific indeed, but does this proliferation lead to its own problems?

As citizen science grows, so too do its logistical challenges. These challenges can take on many forms, ranging from the duplication of particular projects to the difficulties of volunteering across a number of initiatives. Participants have questions about who owns the data they produce, whether they'll be able to shape how it's used into the future, and how their work might have unintended impacts. Indeed, to this stage, the proliferation of citizen science has raised more questions than it has answered, leading to a rich trove of questions for academic and practitioner research.

Some of the most immediate challenges to sort out are the logistical ones: how participants can efficiently engage in multiple projects, maintain ownership of their data, and gain recognition for the skills and experience they're developing. According to informal SciStarter surveys, for instance, half of citizen scientists were engaged with multiple projects, and a plurality of those were involved in projects on significantly different topics. One such solution is that of the SciStarter Dashboard, which, thanks to a grant from the National Science Foundation, provides a centralized location for people to join multiple projects and track their efforts. The Dashboard also provides improved tools for defining, articulating, and sharing the scope of particular projects (and, soon, particular instruments), making it easier for interested members of the public to find and join new opportunities; produce evidence of participation, which may lead to greater validation; and even rate and review related citizen science tools and instruments, much like they rate and review projects. SciStarter is rapidly becoming a Match.com-meets-Amazon for citizen science! Now *that's* mainstream.

Platforms like these provide opportunities for more easily recruiting new participants, enabling participants to engage in new and unexpected projects, and helping individuals demonstrate the value and impact of their time spent on citizen science efforts. What might it look like if participants are accredited for their experiences? Will that shift the balance of power even more?

The coming decade is also likely to bring notable government and legal advances in terms of the practice of citizen science. While it's likely that such questions—like resolving the liability for citizen scientists, protecting personal information, and ensuring open access to research—will be resolved in vastly different ways across different countries, even basic steps towards legal clarity are likely to open opportunities for new projects to be launched by currently hesitant actors.

With larger and more numerous citizen science projects also comes a growing share of responsibilities for all parties, including maintaining the integrity, accessibility, transparency, and accountability of both the science and the participation. Indeed, as we collectively create more and more "participatory" citizen science projects, we should not ignore the more difficult but equally critical task of developing "democratic" citizen science. This means grappling, for instance, with heavy questions about which groups should have the say on which issues. What kinds of democratic decision making should we continue to embrace, particularly when it leads to slow, painstaking action on important issues like climate change? How should we consider those who simultaneously function as passionate citizen scientists and policy advocates alike — or scientists who play both roles? And how do we incorporate and mediate the voice of the public that isn't engaged in citizen science or scientific outreach?

Questions of citizen science like these are intrinsically connected to questions of public participation. In rethinking the citizen/science relationship, what lessons must we apply to the relationships between citizens, and between citizens and their governments? How do we think about core democratic concepts like representativeness, diversity, and public values? And how do we train the public to be productively engaged with decision- and policy-making processes?

Citizen forums, like those hosted by the Expert & Citizen Assessment of Science and Technology (ECAST) network using a process known as "participatory technology assessment," offer one method of enhancing public deliberations. The distributed institutional network consisting of academia, informal science educators and non-partisan policy partners is positioned external to the science and policy development process. As an "honest broker," it is not susceptible to funding and political pressures, as was the case with the former Congressional Office of Technology Assessment (OTA). The participatory technology assessment process engages demographically and geographically diverse lay citizens in an informed, structured deliberation designed to generate public value input for policy and science. It provides a pathway for public concerns that may be poorly expressed through conventional political avenues.

Figure 1: Potential Places for Citizen Science in Decision Making

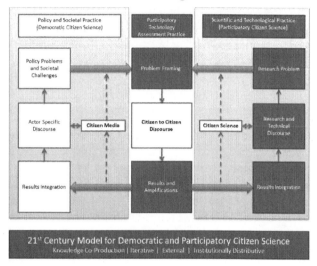

Source: Farooque & Cavalier (forthcoming).

Citizen scientists are not currently part of the design of most participatory technology assessment processes, but they very well could be. As illustrated in Figure 1, citizen media and citizen science can collectively provide input for framing the research, and also utilize the output to initiate citizen media and citizen science projects. The two can collectively identify, report, and act on the data, information, public value, and public policy gaps. From there, it is not difficult to imagine a situation in which citizen science and citizen media work in concert with citizen interests (as envisioned in Lily Bui's chapter). This could create a self-sustaining and iterative process of continual engagement with science, elevating and strengthening both the "participatory" and the "democratic" aspirations of citizen science.

In Closing...

In its richest form, citizen science has the power to transform science and society. Rather than simply recruiting volunteers or producing cool new tools, citizen science reshapes central notions of science and power: the roles of experts and the public, the accessibility of tools and data, and the kinds of questions that are worth asking.

This movement is taking off, and its future isn't guaranteed to be entirely smooth or predictable. The direction it takes over the coming decades will, fittingly, be determined by developments from nearly all sectors. Government agencies will be called on to respond to the will of the people while navigating global research priorities and policies. Industry and scientists will be forced to grapple with intellectual property considerations when their core customers demand more access to data and open tools. And people of all kinds will have the potential to shape entire fields of research and related policies, both globally and in their communities.

The final challenge of this book, therefore, is simple: to treat the volume as the beginning of an exploration into citizen science, rather than a conclusive end. Whether using a resource like SciStarter—to find and join projects, track your contributions, and access the instruments needed to do these projects—as an on-ramp to your adventure, or through more reading and research, we encourage you, the reader, to continue to explore this field and get your own hands dirty with citizen science. Citizen science is, in many ways, a canvas with much open space remaining to be filled. And it's a movement that will undoubtedly shape your life, from the development of health sensors to the data gathered about pollutants in your own backyard, from your role in setting research agendas and related policies to creating new research tools. Now is the time for you to explore the world of citizen science.

ABOUT THE AUTHORS

Lily Bui
Lily Bui is an M.S. candidate for MIT's Comparative Media Studies and an incoming Ph.D. candidate for MIT's Department of Urban Studies and Planning. Her interests lay at the intersection of sensor technology, urban design, participatory science, and public media.

Darlene Cavalier
Darlene Cavalier is a Professor of Practice at Arizona State University's Center for Engagement and Training, part of the School for the Future of Innovation in Society. She is the founder of SciStarter, founder of Science Cheerleader, and cofounder of ECAST, Expert and Citizen Assessment of Science and Technology. She is a founding board member of the Citizen Science Association, a senior advisor at *Discover Magazine*, and a member of the EPA's National Advisory Council for Environmental Policy and Technology. She resides in Philadelphia, PA, with her husband and their four children.

David Coil
David Coil is a Project Scientist in the lab of Jonathan Eisen at the University of California Davis. His background is in microbiology and his current research interests focus on bacterial genomics and microbial ecology. He loves teaching, mentoring, citizen science, and other forms of science communication, including designing an educational board game called *Gut Check: The Microbiome Game*.

Caren B. Cooper

Caren Cooper, Ph.D., is an Associate Professor in Forestry and Environmental Resources at North Carolina State University. Her research focuses on promoting community participation in avian research for stewardship and pollution biomonitoring and mapping. For 14 years, she helped design citizen science projects at the Cornell Lab of Ornithology, including My Yard Counts, YardMap, NestWatch, Celebrate Urban Birds, and the House Sparrow Project.

Robert R. Dunn

Rob Dunn is an ecologist and author, most recently of the book *The Man Who Touched His Own Heart* (2015, Little Brown). His work focuses on using ecological and evolutionary theory to discover organisms of value to society, whether in making beers or in cleaning up industrial waste.

Eric B. Kennedy

Eric Kennedy is a Ph.D. student in the Human and Social Dimensions of Science and Technology program at Arizona State University. His work examines the interface between expertise, public engagement, and decision-making. Kennedy is a 2014 Breakthrough Generation Fellow, and publishes in environmental science, innovation systems, and the social studies of science and knowledge.

Bruce V. Lewenstein

Bruce Lewenstein is a professor in science communication at Cornell University. His research interests focus on studying the ways in which public communication is fundamental to the production of reliable scientific knowledge about the natural world. Alongside extensive work on citizen science, his formal training is in the history of science.

Holly L. Menninger

An entomologist by training, Holly Menninger is a science communicator by passion and practice. She has worked at the intersection of science and society—in policy, natural resource management, and public engagement in science. In 2014, she was named the first director of public science for the College of Sciences at North Carolina State, where she currently oversees a series of initiatives designed to build science literacy beyond NC State's campus.

Gwen Ottinger

Gwen Ottinger is Assistant Professor in the Department of Politics and the Center for Science, Technology, and Society at Drexel University. She is author of *Refining Expertise: How Responsible Engineers Subvert Environmental Justice Challenges*, and is currently working on a National Science Foundation-funded project on the history of community-based air monitoring.

Alex Pang

Alex Soojung-Kim Pang lives in Silicon Valley and is the author of *The Distraction Addiction: Getting the Information You Need and the Communication You Want, Without Enraging Your Family, Annoying Your Colleagues, and Destroying Your Soul*. Pang received a Ph.D. in the history of science from the University of Pennsylvania.

ACKNOWLEDGEMENTS

Producing a book is a massive undertaking, and doubly so when working on such a large and rapidly evolving topic. As such, the project wouldn't have been successful without the significant contributions of a great number of colleagues and friends.

First and foremost, we owe the contributing authors a debt of gratitude. The book writing and publishing process is a challenging combination of both marathon and sprint, demanding patience and energy alike from our collaborators. Moreover, we've been exceptionally fortunate to have contributors who have challenged us intellectually while remaining friendly, helpful, timely, and responsive to us and the volume. We're forever grateful to Lily Bui, David Coil, Caren Cooper, Robert Dunn, Bruce Lewenstein, Holly Menninger, Gwen Ottinger, and Alex Pang and for their friendship, contributions, and assistance, both within and beyond this volume.

An incredible part of this process has been working within a series that embraces rapidly moving subjects, approachable and affordable scientific communication, and emerging authors. Such an enterprise — *The Rightful Place of Science* — requires an editor with experience, tenacity, and determination over a long haul — all qualities of the series editor, G. Pascal Zachary. His commitment to and enthusiasm for the series, this volume, and his colleagues has never waned. Thanks also to Jason Lloyd at the Consortium for Science, Policy & Outcomes for ushering this volume through publication.

For Darlene Cavalier, the length of this volume would double if I were to list all the people who have provided guidance and inspiration, but I will name just a few. David Guston, Ira Bennett, and Mahmud Farooque at Arizona State University's Consortium for Science, Policy & Outcomes and the Center for Engagement in Training in Science and Society have long supported me and my unorthodox approaches. I am so fortunate to join their ranks at ASU, where an entire island of misfits like me have been united, nurtured, and encouraged to think and act big at the new School for the Future of Innovation in Society. Colleagues at the Citizen Science Association, particularly Jennifer Shirk, Rick Bonney, and Greg Newman, coupled with SciStarter team members, advisors, and other leaders in the field, including but certainly not limited to Daniel Arbuckle, Dennis Murphy, Caren Cooper, Catherine Hoffman, Arvind Suresh, Steve Gano, Carolyn Graybeal, Jonathan Brier, Hined Rafeh, Jill Nugent, Eva Lewandowski, David Sittenfeld, Anne Bowser, Leah Shanley, Jenn Gustetic, and Pietro Michelucci, are moving this field forward with a collective impact. We've only scratched the surface in this volume and I encourage readers to follow their work for richer insights. The remarkable Science Cheerleaders for recruiting thousands of new citizen scientists. And there are my personal cheerleaders who have made it possible for me to take some risks: Marc Zabludoff, Corey Powell, Susan Mason, Susan Singer, Bob Russell, Russ Campbell, Richard Sclove, and Bart Leahy. Lastly, a special nod to my husband (Bob), four children (Rebecca, Ronnie, Will, and Teddy), mom (Faith Quinn), and in-laws (Ron and Bonnie Cavalier), who have endured so many years of endless chatter about citizen science that they could probably write their own book on this subject.

For Eric Kennedy, this volume wouldn't have been possible without the support of many academic colleagues and mentors. Arizona State University's Consortium for Science, Policy & Outcomes and School for the Future of Innovation in Society—my academic homes—are second to none. Few academic units would be so encouraging of a graduate student taking on the co-editorship of a book, or of these kinds of projects that focus on public engagement and scholarship outside of the academy. In particular, I owe a debt of gratitude to my advisor of unending patience and cutting insight, Daniel Sarewitz, as well as several colleagues, mentors, and friends who have helped me think more clearly and rigorously about these topics: Ira Bennett, Mahmud Farooque, David Guston, Clark Miller, and Jamey Wetmore at Arizona State, and Ed Jernigan and Kathryn Plaisance at the University of Waterloo, among many others. Most of all, my never-ending thanks to Jenna Vikse for her thoughtful and perceptive insight throughout the process, for pushing me to think more clearly and robustly about my research, and for her limitless support of my every endeavor.

Gwen Ottinger wishes to thank Ruth Breech for her patience with many questions about the making of *Warning Signs*.

Finally, we are grateful for the efforts of everyone involved in the citizen science community. Whether volunteers, project organizers, government decision makers, the innovators who are creating instruments and tools for citizen scientists, or critics, your time and thoughtful efforts have laid the foundation on which projects like this are possible. We hope that this volume serves as a helpful resource, and that we have encouraged new voices to join your ranks.

Made in the USA
Middletown, DE
14 March 2017